REMAKING THE GLOBAL ECONOMY

ECONOMIC-GEOGRAPHICAL PERSPECTIVES

REMAKING THE GLOBAL ECONOMY

ECONOMIC-GEOGRAPHICAL PERSPECTIVES

Edited by
Jamie Peck and Henry Wai-chung Yeung

SAGE Publications
London • Thousand Oaks • Delhi

SAGE Publications Ltd
6 Bonhill Street
London EC2A 4PU

SAGE Publications Inc.
2455 Teller Road
Thousand Oaks, California 91320

Sage Publications India Pvt Ltd
B-42, Panchsheel Enclave
Post Box 4109
New Delhi 100 017

British Library Cataloguing Publication data

A catalogue record for this book is available from the British Library

ISBN 0 7619 4897 X
ISBN 0 76 19 4898 (pbk)

Library of Congress Control Number available

Printed in India by Gopsons Papers Ltd, Noida

CONTENTS

INTRODUCTION

PART ONE   GROUNDING GLOBAL FLOWS

LIST OF TABLES

LIST OF FIGURES

NOTES ON CONTRIBUTORS

Ash Amin is Professor of Geography at the University of Durham, UK. His most recent publications include: *Cities for the Many not the Few*, Policy Press, 2000 (with Doreen Massey and Nigel Thrift), *Cities*, Polity Press, 2002 (with Nigel Thrift), *Placing the Social Economy*, Routledge, 2002 (with Angus Cameron and Ray Hudson), *Cultural Economy: A Reader*, Blackwell, 2003 (edited with Nigel Thrift), *Organisational Learning: From Competences to Communities*, Oxford University Press, 2003 (with Patrick Cohendet).

Neil Brenner is Assistant Professor of Sociology and Metropolitan Studies at New York University, USA. He is co-editor, with Nik Theodore, of *Spaces of Neoliberalism: Urban Restructuring in Western Europe and North America* (Blackwell, 2002). His research and teaching focus on critical urban studies, state theory and socio-spatial theory.

Neil M. Coe is Lecturer in the School of Geography, University of Manchester, UK. His main research interests within economic geography include the dynamics of cultural industries and the service sector, processes of internationalization and globalization, and transnational corporate activity. He has published several papers on the software and computer service industries of the UK, Ireland and Singapore. His current research focuses on transnational IT-sector linkages between Southeast Asia and the US, and the internationalization of retailing.

Michael Conroy is Program Officer at the Ford Foundation, USA.

Peter Dicken is Professor of Geography at the University of Manchester, UK. His major research interests are in global economic change and transnational corporations. The Fourth Edition of his book on the global economy, *Global Shift*, is published in 2003 by Sage.

Meric S. Gertler is Professor of Geography and Planning, and holder of the Goldring Chair in Canadian Studies at the University of Toronto, Canada. He also co-directs the Program on Globalization and Regional Innovation Systems at the University of Toronto's Centre for International Studies, where he carries out research on the role of institutions and social context in the shaping of corporate practices in North America and Germany. His publications include *The New Industrial Geography* (with Trevor Barnes, Routledge, 1999), *The Oxford Handbook of*

Economic Geography (with Gordon Clark and Maryann Feldman, Oxford University Press, 2000), *Innovation and Social Learning* (with David Wolfe, Palgrave/Macmillan, 2002) and *Manufacturing Culture: The Gover-nance of Industrial Practice* (Oxford University Press, 2003).

Amy Glasmeier is Professor of Geography at the Pennsylvania State University, USA.

Ray Hudson is Professor of Geography and Chair of the International Centre for Regional Regeneration and Development Studies at the University of Durham, UK. His recent publications include *Producing Places* (Guilford, 2001) and *Digging Up Trouble: The Environment, Protest and Opencast Coal Mining* (Rivers Oram, 2000, with Huw Beynon and Andrew Cox). His research interests are in geographies of economies, political economies of uneven development, and issues of territorial development, especi-ally in the context of Europe. Current research includes work on the socio-spatial transformation of the former UK coalfields and on the links between corporate restructuring and regional development strategies in Europe.

Philip F. Kelly is Assistant Professor of Geography at York University, Toronto, Canada. With research interests in the political economy of Southeast Asian develop-ment, he is currently examining the linkages between Filipino transnational migration and local labour market processes. He is the author of *Landscapes of Globalization: Human Geographies of Economic Change in the Philippines* (Routledge, 2000) and co-editor of *Globalisation and the Asia Pacific: Contested Territories* (Routledge, 1999).

Roger Lee is Professor of Geography and Head of Department of Geography at Queen Mary, University of London, UK. He is the co-editor (with Jane Wills) of *Geographies of Economies* (Arnold, 1997) and (with Andrew Leyshon and Colin Williams) of *Altern-ative Economic Geographies* (Sage, 2003). His research interests are in the social construction of economic geographies and the role of money and the social relations of capitalism in linking the personal and the global and so constraining and enabling such social construction and the possibilities of proliferative geographies.

Anders Malmberg is Professor in the Department of Social and Economic Geography, Uppsala University, Sweden. His research interests cover the relation between industrial change and regional economic development, with particular focus on spatial clustering and localized processes of innovation and learning. He has published a number of papers on these issues in recent years, and is the co-author of *Competitiveness, Localised Learn-ing and Regional Development* (Routledge, 1998).

Kris Olds is Associate Professor in the Department of Geography at the University of Wisconsin-Madison, USA. His recent publications include (as co-editor) *Globalisation and the Asia-Pacific: Contested Territories* (Routledge,

1999), *Globalisation of Chinese Business Firms* (Macmillan, 2000), and (as author) *Globalization and Urban Change: Capital, Culture and Pacific Rim Mega-Projects* (Oxford University Press, 2001). He was based at the National University of Singapore from 1997 to 2001, and has a PhD in Geography from the University of Bristol (1996). His research currently focuses on global city formation processes in Pacific Asia, urban redevelopment processes, global economic networks, and transnational communities.

Jamie Peck is Professor of Geography and Sociology at the University of Wisconsin-Madison, USA. He is author of *Work-Place: The Social Regulation of Labor Markets* (Guilford, 1996) and *Workfare States* (Guilford, 2001) and co-editor (with Kevin Ward) of *City of Revolution: Restructuring Manchester* (Manchester University Press, 2002). With re-search interests in economic regulation and governance, labour markets and urban polit-ical economy, he is currently working on a study of contingent labour in the US.

Erica Schoenberger is Professor of Geography at the Johns Hopkins University, USA.

Nigel Thrift is Professor of Human Geography at the University of Bristol, UK.

Adam Tickell is Professor of Human Geography at the University of Bristol and has previously lectured at the universities of Leeds, Manchester and Southampton. He is editor of *Transactions of the Institute of British Geographers* and review editor of the *Journal of Economic Geography*. His work explores the geographies and politics of inter-national financial reform, governance structures in the UK, and the reconfiguration of the political commonsense.

Henry Wai-chung Yeung is Associate Professor in the Department of Geography, National University of Singapore. His research interests cover broadly theories and the geography of transnational corporations, Asian firms and their overseas operations and Chinese business networks in the Asia-Pacific region. He is the author of *Transnational Corporations and Business Networks* (Routledge, 1998), *Entrepreneurship and the Internationalisation of Asian Firms* (Edward Elgar, 2002), and *Chinese Capitalism in a Global Era* (Routledge, 2003). He is also the editor of *The Globalisation of Business Firms from Emerging Markets*, two volumes (Edward Elgar, 1999) and co-editor of *Globalisation and the Asia Pacific* (Routledge, 1999) and *The Globalisation of Chinese Business Firms* (Macmillan/Palgrave, 2000).

PREFACE

This collection has been assembled to mark the many contributions that economic geographer Peter Dicken has made in a professional career spanning three-and-a-half decades. It is testimony to the widespread respect and affection for Peter that no-one we approached about contributing to this book, despite their busy schedules, hesitated before saying yes. Well, this is not exactly true. There was one contributor who was at first somewhat reticent – Peter himself. He has never been the one to blow his own trumpet and was understandably unsettled at the prospect of the two of us blowing it for him. But it soon became clear that none of us had any interest in producing a book that was retrospective or introspective, even if Peter deserves a little hagiography. Reflecting Peter's own approach, we wanted the book to look forward and outward, to take stock of what economic geographers have contributed to the 'globalization debate' and to explore new frontiers in this vibrant field of interdisciplinary engagement. And the book would also, we hoped, demonstrate some of the range and depth of what economic geographers can bring to the table in globalization studies. With contributions from Asia, North America and western Europe, the key issues explored in *Remaking the Global Economy* include the globalization of firms, people and capital (Part One), organizational learning and business knowledge, industrial districts and innovation systems (Part Two), and ideologies of neoliberal globalization and interactions between firms, regions and nation states (Part Three). The book therefore seeks to engage with some of the fundamental strands of the globalization debate, building upon Peter Dicken's compelling insight that globalization must be understood as an ongoing *geographical* project.

Today, the global economy is more complex and interdependent than ever before, for all the important historical continuities. The study of globalization is now firmly on the agenda across the social sciences. Economic geographers have developed distinctive perspectives on the globalization process, eschewing 'flat-earth' visions of homogenization and convergence in favour of more nuanced treatments of globalization as an uneven, differentiated and dynamic process. Globalization, in other words, has a geography and it is a geography that is on the move. This is where the chapters collected in this volume make their contribution. Notwithstanding the book's roots in economic geography, each chapter connects to and further develops interdisciplinary insights into the complex process of economic globalization and its impact on the spatial organization of firms, markets, industries, institutions and regions.

The book is organized into three parts. Part One explores some of the fundamental ways in which global flows can be considered to be 'grounded'; Part Two unpacks the spatiality of global knowledge and learning; and Part Three analyses the reconfiguration of global rule regimes and their implications for territorial development. As a prelude to this, Chapter 1 assesses the contributions of Peter Dicken's work in relation to the interdisciplinary field of 'globalization studies'. Here, our aim is to outline how this particular economic-geographical perspective has enriched understandings of patterns and processes of globalization, together with the attendant ways in which global economic relations have been remade. Through the lens of Dicken's work, we also seek to identify some insights into how economic-geographical perspectives might be better integrated with the evolving field of 'globalization studies' in cognizant disciplines like global political economy, international economics, strategic management and international business studies. The chapter therefore opens up some analytical 'problematics' to be followed up by subsequent chapters.

One of the most intractable problems for contemporary studies of global economic change is the question of what exactly *flows* across territories and places in the form of globalization processes. The four chapters in Part One of this volume examine how these flows are geographically constituted, exploring the 'groundedness' of global flows of *firms*, *people* and *capital*. The spatiality and territorialization of globalization processes typically evades the analytical attention of most social scientists, though for many geographers these have been amongst the most important issues in play. In unpacking the spatiality of globalization, geographers have contributed to the understanding of how global flows are grounded in specific places, regions and territories. This grounding of global flows is important in economic-geographical studies for two reasons. First, without an appropriate appreciation *where* globalization takes place (literally as well as metaphorically), a key explanatory dimension of globalization – as a set of tendencies or processes that (unevenly) bring together distant and disparate locales within an increasingly interdependent world – is missed. Second, the resultant networks of global connections should not be conceived as just 'hanging in the air', constituting an exterior and superordinate force to which localities and regions must respond. In some accounts, for example, globalization is caricatured as a highly abstract system of 'flows' operating across boundaries, places being reduced to mere nodes in these floating network-systems. Absent from this kind of sociological topography of globalization is an appreciation of the complex ways in which such flows, networks and connections operate unevenly across space, and how they 'touch base' to bring about growth, prosperity and development for some places, while marginalizing others.

In this sense, the four chapters in Part One provide a much-needed discussion of how global flows are grounded in historically and geographically specific formations. For Peter Dicken in Chapter 2, the 'global' corporation is somewhat less global than the name suggests because, despite their rapidly expanding foreign activities and investments, many of today's largest transnational corporations

(TNCs) are indeed grounded in specific places. In fact, 'global' corporations turn out to be quite deeply embedded in their home economies, culturally and economically. Dicken's analysis problematizes the globality of 'global' corporations, separating a complex reality from the business- and media-driven hyperbole. His concern to reveal the hidden geographies of globalization dovetails with those of Neil Coe, Philip Kelly and Kris Olds who, in Chapter 3, seek to make sense of the deepening 'cross-border' flows of people in relation to the production of transnational economic spaces. For them, the movement of people and the attendant reorganization of social networks across geographic space not only challenges the conventional, capital-centric view of globalization, but also opens up new horizons for analysing the groundedness of global flows of labour, expertise and social networks. Through two case studies of transnational flows of people grounded in the property and the information technology sectors, their chapter shows that globalization and transnationalism are different facets of the same profoundly geographical restructuring of economic activity.

An equally valid analytical strategy, of course, would focus on what might be thought of as the newly constituted 'remote' spaces in a globalizing world – the *marginalized spaces*. Marginalization is the other side of the geographical coin to the hypertrophied 'over-inclusion' of places like world cities and global financial centres. In globalization studies, the *sine qua non* for economic globalization is the hyper-mobility of financial capital. Paradoxically though, this very hyper-mobility is supported by a global financial architecture that is constructed around 'strong nodes' like global financial centres. The process of financial globalization, however, cannot be reduced to stories of Tokyo, London and New York, because exclusion and marginalization are just as much parts of this process. The two chapters by Roger Lee and Erica Schoenberger speak to these concerns by virtue of their analytical and empirical focus on marginalized and 'emergent' spaces within global flows of capital. They use different readings of actually existing globalization processes to raise searching questions about the sustainability of these processes. In Chapter 4, Lee analyses how the globalization of financial capital hinges on the construction of uneven development, producing what he terms the 'marginalization of everywhere'. Using emerging markets as the central theme, he demonstrates how certain economies and places are purposefully marginalized in the global (re)switching of different circuits of capital. The following contribution from Erica Schoenberger in Chapter 5 also explores some of the neglected spaces of globalization, tracing the emergence of a new field of international direct investment in environmental management. Through the case of a French TNC in the water and waste treatment businesses, she analyses how the globalization of environment management represents a key moment in capital's restless search for spatial and institutional 'fixes' in the face of inherent crisis tendencies. Again, this 'new frontier' for global corporations exhibits a very particular geography of unevenness, exclusion and concentration.

These close tracings of the globalization process at work stand in sharp contrast to the polarized accounts found in much of the globalization literature. Too often, this takes the caricatured form of a contest between two (apparently irreconcilable) positions – 'everything has changed, thanks to globalization' versus 'nothing has changed, we've been there before'. Geographers have tended to cut a different path through the globalization debate. They have been rather less concerned with measuring quantitative change in the global economy, narrowly conceived, than with the causative foundations underlying these changes and the altered qualitative relations – especially between places – that they entail. The influential body of work that has been produced on technology, learning and knowledge is a prime illustration of such concerns. The transposition of technologies across space remains highly problematical, not least because the spatial transfer of knowledge and practices is inherently connected to the social organization of these technologies and their variable embeddedness in networks, organizational practices and places. Modes of social learning and technology transfer are constituted both geographically and through network relations.

The four chapters in Part Two critically assess the spatiality of global learning and knowledge from economic-geographical perspectives. In Chapter 6, Meric Gertler contributes to the recent debate around the convergence-divergence of different 'varieties of capitalism' in the global economy. He reveals how the processes of cross-border learning and practice are shaped by the cultural and institutional specificities of different places. Supporting Dicken's arguments in Chapter 2, Gertler concludes that the spatial life of learning and knowledge significantly constrains the 'globality' of firms. In a parallel fashion, Ash Amin in Chapter 7 explores the spatiality of tacit learning in distributed organizations, identifying qualitatively differentiated spaces of corporate learning within the globalizing economy. Focusing on the everyday interactions in globalizing organizations, he questions the deterministic view of spatial proximity in organizational learning and proposes, instead, a more relational view of the interaction between space and learning. Here, he shows analytically how organizational learning occurs through the emergence of different 'communities of practice', drawing on spatially 'stretched' connectivities. On the one hand, relational proximity makes it more difficult to ground global connections because it exists in organizational rather than physical spaces. On the other hand, however, different spaces of tacit knowledge and practices can be brought together through relational proximity among actors.

In Chapter 8, Nigel Thrift takes on some of these claims on the nature of everyday learning and knowledge, drawing attention to what he sees as a proliferation of new practices of capitalist power – what he calls the might of 'might'. He contends that new forms of creativity and standardization enable new possibilities for firms and organizations to (re)engineer space and time in the service of greater returns and profitability. Focusing on different circuits of spatial and temporal knowledge, he theorizes how we might think of global knowledge in radically different spatial and temporal terms. In Chapter 9, Anders Malmberg examines one peculiar form of spatial arrangement, clusters, in order to demonstrate the importance of both local milieus and global connections in the processes of social learning and knowledge transfer. In recent years, geographers have (re)discovered several spaces in which globalization processes appear to touch base. Since Alfred Weber and Alfred Marshall, agglomeration has occupied a special place in the nomenclature of economic geography (and, more recently, in the 'geographical economics' of Paul Krugman and Michael Porter). The growth and development of industrial clusters and similar forms of concentrated territorial development seems to reaffirm the role of agglomeration economies. The variegated capacities of recently emergent agglomerations in 'holding down' global flows, however, have only been explored in rather stylized terms. Malmberg insists on the need for greater clarity in the analysis of cluster dynamics, especially relating to the global connections of actors in these clusters.

Economic globalization is clearly not just about a set of material processes operating across national boundaries. As Part Three of the volume demonstrates, these processes are also located in the *ideological* realm, since 'globalization' is in part a political project focused on the reconfiguration of global rules and regimes. Countering the pervasive conception of globalization as a triumphal 'end state' of market capitalism, a wide range of social theorists have contended that processes of globalization must instead be understood to be politically mediated, socially structured and discursively framed. Political-economic discourses relating to issues like the evisceration of the state, the imperatives of labour-market flexibility and trade union 'realism', or the necessity of ongoing liberalization in trade and financial markets, even if they are presented as naturalized 'facts of life', do not spring automatically from some underlying economic logic. Instead they must be understood to be socially produced discourses that reflect, serve and help realize political-economic interests. By insistently questioning the politics of globalization, geographers have helped to specify the causal agency of globalization processes, their inescapably political construction and their variable concrete manifestations.

In Chapter 10, Adam Tickell and Jamie Peck explore the theoretical and political status of neoliberalism in remaking the global economy. The ascendancy of neoliberal ideologies in the course of the last three decades has shadowed the intensification of 'real' globalization processes in the economic realm, such that ultimate causality and logic priority are difficult to determine in any kind of unambiguous way. Orthodox globalization narratives and neoliberal political discourses both tend to privilege 'market rule', presenting this as a self-evident and practically irresistible future. In a sense, then, neoliberalism and globalism are mutually naturalizing discourses. There is a need to deconstruct both of these discourses in the context of contemporary political-economic conditions. In reality, neither globalism nor neoliberalism are as totalizing and monolithic as they may seem at first, despite the fact the effects of both are undeniably pervasive. Tickell and Peck make the case for a close interrogation of the neoliberal political project in a way that is sensitive to its very uneven geographies and its complex evolution over time. They also contend that the ideological dynamics of the *process* of neoliberalization cannot be reduced to the aggregate effects of merely 'local' political agency (as if, say, Thatcherism + Reaganomics … = neoliberalism), but must be traced out in ways that are attentive both to the 'generic' character of neoliberalism and its local manifestations.

These issues relating to the remaking of the global economy through neoliberal globalism are further explored and interrogated in the following three chapters. In Chapter 11, Amy Glasmeier and Michael Conroy trace the evolution and governance of the emerging global trade regime. They make a powerful argument for an economic-geographical perspective on the tradeoffs and effects of the trade regime, which tends to exclude most developing countries from the potential upsides of globalization. More specifically, they examine the contested legitimacy of the World Trade Organization, the high-handedness of the US Trade Representative's Office and the over-reaching claims of wealthy countries concerning the use of natural resources and the management of intellectual property rights. These are the kinds of policy questions on which economic geographers can, and should, be making a mark. In making such a mark, a premium will be placed on the kinds of grounded knowledges of the global that are a key feature of economic-geographical contributions.

The last two chapters by Neil Brenner and Ray Hudson focus on ongoing development in urban governance and production systems in an integrating Europe. In Chapter 12, Brenner examines urban entrepreneurialism in western Europe in relation to the ongoing processes of economic globalization, European integration and the crisis of the Keynesian welfare national state. He draws upon Bob Jessop's strategic-relational approach to develop a spatialized state theory, arguing that recent transformations reflect a deepening neoliberalization of urban politics. Brenner also underscores an important exception to the claims of ultra-globalists: nation-states in western Europe are not 'captives' of globalization processes; they are key institutional players in these processes. The way in which governmental institutions enact globalization processes – at scales ranging from the urban to the supranational – is represented here as a new form of 'state spatial strategy'. Many of these observations on urban transformations in western Europe are echoed in Ray Hudson's analysis of global production systems and European integration. In Chapter 13, Hudson argues that rapid transformations in production systems within Europe are as much outcomes of corporate strategies, orchestrated from the headquarters of 'global' corporations, as they are consequences of changing political-economic circumstances of Europe. His analysis shows once again that firms and states are very much active agents in remaking the global economy; they are certainly not merely passive actors in, nor are they simply victims of, abstracted globalization processes. By the same token, the process of European integration is not producing a flattened economic landscape in which equilibrating forces hold sway, but on the contrary, is generating new forms of uneven spatial developments and new landscapes of political-economic power.

The contributions collected here, then, underline the distinctiveness of economic-geographical perspectives on the globalization process, a field that has been profoundly shaped by Peter Dicken's work. We hope that *Remaking the Global Economy* will stand as one of the many markers of Peter's conspicuous contributions, while also opening up new terrains for spatialized globalization studies. On a more personal note, our editor at Sage, Robert Rojek, deserves special thanks for his faith in this project, which at times must have sorely been tested by us. And we are grateful also to our contributors for enduring the torrent of emails and for always responding so positively. Nick Scarle has diligently drawn some of the figures at very short notice. Finally, and on behalf of the contributors as well as ourselves, we would like to think that this volume represents a modest down-payment on the debts – both personal and professional – that we owe Peter. There will always be a little bit of Manchester in the two of us, and more than anyone it was Peter who put it there. As well as teaching us how to do economic geography, he showed us how to enjoy it at the same time. And it is thanks to Peter that we still both wince every time we see a split infinitive (If there are any here, it is because we left them in on purpose.) He has been his usual, quietly-supportive self during the production of this book, even though there was a sense in which the very thought of it seemed to unnerve him. In a sense, though, this made the project even more enjoyable. We always said that Peter couldn't retire until this book came out, which just added to the list of excuses we were accumulating for delaying publication. Now it's out, it is a nice thought that maybe he'll take it with him on that first, symbolic walk to the Post Office.

Jamie Peck and Henry Yeung
Madison and Singapore/Manchester

INTRODUCTION

Chapter 1

MAKING GLOBAL CONNECTIONS:
A GEOGRAPHER'S PERSPECTIVE

Henry Wai-chung Yeung and Jamie Peck

FROM SPATIAL ORGANIZATION TO *GLOBAL SHIFT*:
CONTEXTUALIZING THE WORK OF PETER DICKEN

The study of global economic transformation has become a central concern to virtually all disciplines in the social sciences and beyond. Economic geography has played an important role in these conversations and there have been few more important voices than that of Peter Dicken. Peter can be credited with putting globalization on the agenda in economic geography, while also serving as one of the discipline's most influential advocates in fields like international economics, global political economy and international business studies. And while he has been an intellectual pioneer, he has never been content to sit still. Echoing Peter's approach, this book seeks to push forward and look beyond existing frontiers, though we have allowed ourselves one exception to this rule. This chapter underlines the central motive for the volume by sketching a genealogy of Dicken's work. We present a brief historical survey of key research themes in the debate around global economic transformation in the period since the 1970s. We identify some of Dicken's key contributions in economic geography and the geography of international business and assess these in relation to how major debates have been moving in cognate disciplines. Our primary focus is on assessing Dicken's research impact on the ways in which we think about the global economy. In this context, we will not dwell on his significant contributions to teaching and to the academic profession more generally.[1] Interrogating key issues in the globalization debate through the lens of his work might be seen by some as a partial approach, but what is striking in the examination that follows is just how many of the critical questions in this debate were foreshadowed by his work.

Dicken tackles globalization research in a distinctive way. His work is measured and careful but not unnecessarily cautious; it is authoritative but not declaratory; and it combines a grasp of complexity with a parallel insistence on clarity. In substantive terms, his work does not re-circulate 'flat-earth' visions of globalization as an homogenizing force, but instead tenaciously interrogates a complex set of transformative processes. And while it is, perhaps, an occupational hazard in the field of globalization studies to succumb to hyperbole and exaggeration, this is another reason why his work stands out. Globalization, the epitome of a 'big picture' issue, is handled here with subtlety and dexterity. Dicken's work speaks to the globalization debate, therefore, in distinctive ways – stylistically as well as substantively. Our focus on his signal contributions to this broadly based and diverse project is consequently unapologetic, for it opens up a range of issues around the conceptual formation of what we might now call 'globalization studies' and the associated interdisciplinary research agenda.

At some risk of oversimplification and caricature, we discuss the main themes and contributions of his published work with reference to four sequential phases beginning in the early 1970s:

1   spatial decision-making, external control and regional development;

2   global shifts, transnational corporations and industrial change;

3   firms, states and global networks; and

4   global production networks, territorial organization and relational analysis.

We focus on the underlying conceptual arguments, frameworks and apparatuses that have been developed in Dicken's work. While a notable feature of these contributions is the way in which central themes and contentions are supported and illustrated through detailed empirical investigation, we can only concentrate on some of the central themes and connections, linking them to parallel developments in the interdisciplinary research effort around globalization studies. In a number of important respects, Dicken's work has been ahead of the curve here, but it has also been reciprocally receptive to debates and conceptual insights from a range of social science disciplines. This pathway through the globalization debate is revealing both in terms of the lineage of Dicken's work and as a means of narrating the evolution of the debate itself.

ORIGINS: SPATIAL DECISION-MAKING, EXTERNAL CONTROL AND REGIONAL DEVELOPMENT

Trained as a 'location theorist' under the tutelage of David M. Smith during the mid 1960s, Dicken's early empirical research focused on clothing firms in Manchester, England. During the late 1960s and through the 1970s, his intellectual reputation and disciplinary identity was shaped, in part, by his productive research partnership with Peter Lloyd, including their successful textbook on location-theoretic economic geography (Lloyd and Dicken, 1972; 1977; Dicken and Lloyd, 1990). Significantly, this statement of the theoretical orthodoxy came from two writers who were beginning to see the limitations of neoclassical economic geography. In Dicken's case, this came in the form of an early engagement with behavioural economics and organizational theory. His research contributions during the 1970s marked a significant departure from the prevailing mainstream thinking in economic geography and regional studies.

By the end of the 1950s, neoclassical notions of the firm within economic systems had been seriously challenged by a group of behavioural scientists led by Herbert Simon and James March. Neoclassical assumptions of perfect rationality and information amongst economic actors, the so-called *homo economicus*, seemed increasingly misplaced in a world of bounded rationality and uncertain information. Influenced by these notions of behavioural constraints on human decision-making, Dicken (1971; 1977) began to question some of the central nostrums of neoclassical location theory – the conceptual mainstay of economic geography during the 1960s (e.g. Haggett, 1965). The 'quantitative revolution' in economic geography and regional science had established spatial analysis and methodological individualism as the approaches of choice. Scott (2000: 23) observed that in North America, 'spatial analysis and regional science reached the zenith of their influence some time in the late 1960s and early 1970s'.

Having spent most of the 1969–1971 period in Ontario, Canada, where he continued to work on his survey data on clothing firms in Manchester, Dicken's first major article was published in *Economic Geography* in 1971. Examining aspects of the *spatial decision-making behaviour of firms*, he noted then an 'increasing dissatisfaction with the highly constrained and artificial behavioural component of normative theory' (1971: 426). This article reflected a degree of scepticism concerning the methodological and theoretical conventions of regional science. Instead, Dicken drew on insights from behavioural science and organization studies, especially in systems analysis, organization theory and communications theory. As shown in Table 1.1, this initial intervention was to contribute to economic geography in several ways. First, the article conceptualized firms as open systems 'operating within, and interacting with, an external environment' (Dicken, 1971: 427; emphasis omitted). This suggestive conceptualization of firm-environment interactions within a dynamic system was part of a wider body of work that was opening up new approaches to industrial and enterprise geography (e.g. Hamilton, 1974). And many of these insights remain relevant to today's research initiatives, particularly in strategic management and organization studies, where there is a continuing concern to unpack neoclassical conceptions of the firm as a 'black box' responding purely to price signals and market mechanisms. Distinct echoes of this work can also be seen in the recent resurgence of interest in the firm in economic geography (see Taylor and Asheim, 2001; Yeung, 2001).

Second, the article contributed to a behavioural theory of the firm in the way that it linked decision-making capabilities to internal organizational structures. As

Dicken (1971: 427) argued, '[t]he larger the organization the more complex is its internal structure likely to be and ... this has important decision making implications.

Table 1.1   Main contributions of Peter Dicken to interdisciplinary studies of global economic change

Key Works	Core Concepts	Related Disciplines	Main Impact

1970s
Dicken (1971;
1977)
Dicken (1976)
	• the firm as a system
	• the environment of
	  the firm
	• external control of
	  regions
	• transnational
	  corporations and
	  business strategies
		• behavioural sciences
		• industrial economics
		• organization studies
		• business history
		• regional studies
		• industrial
		  organization
		• development
		  economics
		• strategic
		  management
			• growth of behavioural
			  (economic) geography
			• spatial decision-making
			  behaviour of firms
			• widely cited paper
			  (citation classic)
			• external control of firms
			  and regions less important
			  than its exercise
			• control depends on spatial
			  units and scales of firms
			  and their organizational
			  variables

1980s
Dicken
(1986a;
1990a)
	• global shift
	• global industrial
	  change
	• globalization
	• transnational
	  corporations and
	  their FDI activities
		• industrial economics
		• economic sociology
		• development studies
		• international
		  business
		• strategic
		  management
			• widely cited book (citation
			  classic, 1988§–2002)
			• industrial change to be
			  understood in relation to
			  broader global processes
			• studies of TNCs and
			  foreign direct investments
			  in geography and other
			  social sciences

1990s
Dicken
(1992a;
1998a)
Dicken
(1992b;
1994; 1997;
1998b)
Dicken et al.
(1997a)
Olds et al.
(1999)
Dicken and
Thrift
(1992)
Dicken et al.
(1994)
Dicken and
Miyamachi
(1998)
Dicken and
Yeung
(1999)
	• global shift
	• globalization
	• political economy of
	  firms and states
	• local embeddedness
	  of firms
	• networks and
	  organizations
		• global political
		  economy
		• international
		  economics
		• international
		  business
		• economic sociology
		• strategic
		  management
		• urban and regional
		  studies
			• very widely cited book
			  (quadruple citation classic,
			  1992–2001)
			• *Global Shift* widely
			  recognized as one of the
			  best books written on
			  globalization
			• Paper in *Economic
			  Geography* (1994) well
			  recognized in
			  management (see Kobrin,
			  2001) and reproduced in
			  *Advances in Strategic
			  Management*, 1994.
			• widely cited paper in
			  economic geography
			• first in geography to
			  expound embeddedness
			  and networks

Table 1.1   continued

2000–			

Dicken (2000)
Dicken and
  Malmberg
  (2001)
Dicken et al.
  (2001)
Henderson et
  al. (2002)
	• firm-territory
	  relations
	• global production
	  networks
	• relational
	  perspective on the
	  global economy
		• regional studies
		• studies of innovation
		• strategic
		  management
		• development studies
		• global political
		  economy
		• economic sociology
			• understanding regional
			  development and the
			  territorial aspects of firm
			  activities
			• potential theoretical
			  framework for analysing
			  the global economy

This early investigation into the complex interrelationships between internal organizational structures and corporate decision-making capabilities of firms, not only opened up questions for subsequent research on the geographies of enterprise, but also had clear parallels with developments then in transaction costs economics (e.g. Williamson, 1970; 1975) and strategic management (e.g. Stopford and Wells, 1972).

Third, the analysis of spatial decision-making behaviour of *very large* business organizations in this article foreshadowed an enduring concern with transnational corporations (TNCs) as movers and shapers of global economic change. For example, Dicken (1971: 431) noted that 'most large business organizations possess a highly developed and sophisticated adaptive structure… In some cases, actively searching the environment for new business opportunities may be an established part of the firm's operations.' These early conceptions of organizational capabilities in information gathering/processing and decision-making within large firms found strong parallels in the still embryonic theories of multinational enterprises, in particular Hymer's (1960/1976) market power theory, little known until its posthumous publication in 1976. While Dicken's conception of large business organizations did not precede Vernon's (1966) product life cycle hypothesis of international investment, it certainly anticipated numerous other theorizations of foreign direct investment (FDI): the internalization theory (Buckley and Casson, 1976), the eclectic framework of international production (Dunning, 1977), the information-processing theory (Egelhoff, 1982), and the organizational learning perspective (Barkema and Vermeulen, 1998).

During the second half of the 1970s, Dicken moved on from this early concern with the spatial organization and decision-making behaviour amongst large firms to focus more explicitly on the organizational structures and business strategies of *transnational corporations*. The 1970s witnessed the growing (inter)penetration of different national economies by TNCs via their FDI activities. These processes of corporate internationalization were especially visible in the UK economy, which since the nineteenth century had been 'globalized' in one way or another. And it is hardly far-fetched to say that material conditions in Dicken's 'home base' – the classic industrial city of Manchester – exposed these processes in a particularly vivid form: a city-region that was once on the controlling end of the globalization process, as the 'Cottonopolis' of the nineteenth century, now found itself very much on the receiving end, as a focus for American investment in Europe and, starting in the early 1970s, as a space of deindustrialization and restructuring (see Dicken, 1976; 1980; 2002; Dicken and Lloyd, 1976; 1980). Internationalization was nothing new in this region, but neither – quite clearly – was it a timeless and unchanging process. There were indications, in fact, that the qualitative form of the internationalization process was beginning to change in the 1970s as spatially integrated production complexes were formed and as transnational corporations extended their reach.

In 1970, there were about 7,000 TNCs in the world and more than half of these TNCs were based in the US and the UK (UNCTAD, 1994). By the mid 1970s, FDI was increasing at a faster rate than world trade for the first time, whilst in a subtle way international production increasingly came to resemble international trade, since many countries were beginning to regard TNCs less as a kind of 'threat' and more as a potential means of capturing the upside benefits of the international division of labour and the globalization of markets (Dunning, 1993). Of the numerous competing theories of multinational enterprises and FDI, only Hymer's (1960/1976) market power theory sought to explain the effects of FDI on regional development – the 'external control' of the local/regional economy. In explaining why national firms were involved in international operations, Hymer (1960/1976: 23) argued that the desire to *control* foreign operations was a key motive: '[i]f we wish to explain direct investment, we must explain control'. He explained this trend towards international production by the desire of national firms to gain or maintain their oligopolistic advantages and market power over certain market imperfections abroad.

Situating his research in these theoretical and empirical contexts, Dicken published his highly cited paper on *external control* by multiplant business enterprises and regional development in *Regional Studies* in 1976 (see Table 1.1). In geographical terms, he problematized the 'external' in external control and argued that the key to defining external control lay with understanding 'the actual or potential clash of interests between the goals of multiplant business enterprises, many of which now operate on a global rather than a national scale, and the interests of local communities' (Dicken, 1976: 404). These tensions between managerial control, capital mobility and community interests would later represent one of the central analytical and political motifs of the emergent wave of 'restructuring' studies (e.g. Massey and Meegan, 1979; Bluestone and Harrison, 1982). A further distinctive characteristic of Dicken's work at this time was the way that it problematized the issue of spatial scale, for the operation of TNCs 'across scale' raised the vexing question of the real nature and consequences of 'external' control (see also Brenner and Hudson, in this volume). More specifically, he argued that the size of the spatial unit (regions, nations and so on) is critical to how we define 'external' control.

In organizational terms, Dicken (1976) unpacked the meaning of 'control' in external control. Ownership structures and national origins, for example, would tend to influence both the nature and extent of external control. Similar arguments have been taken up recently in international political economy and organization studies in relation to the issues of divergent capitalism(s) and business systems (see Dicken and Gertler, in this volume). Dicken (1976: 406) observed that 'United

States owned firms tend to be larger, more capital intensive, and more highly concentrated in certain economic sectors than domestic [British] firms'. He further pointed out, however, that to determine the exact extent of control, we need to pay attention to specific business functions (e.g. finance, marketing, and manufacturing) and organizational variables (e.g. size and sector). The latter point about parent-subsidiary relationships as an organizational variable is particularly important in view of the recent resurgence of research interest in subsidiary initiatives in management and international business studies (e.g. Birkinshaw and Hood, 1998) and economic geography (e.g. Phelps and Fuller, 2000; Yeung et al., 2001).

There were also implications for strategic management, given the strong connections between corporate strategy and organizational structure/control (see Bartlett and Ghoshal, 1989; Gupta and Govindarajan, 1991; 2000). In Dicken's (1976: 410) words,

> the impact of multiplant enterprises on geographical space is a function of far more than just the pattern of control, whether potential or actual. Certainly control is important, but it cannot be divorced realistically from the strategy being followed by an enterprise and the structure it has evolved to implement that strategy.

It was during the 1970s, then, that Dicken broke from the neoclassical fold, opened up the black box of the business enterprise and posed new questions about the meaning of 'external control', drawing creatively on behavioural science, industrial economics, and strategic management. His most significant contribution to subsequent studies of global economic change came from the conceptual innovations that connected spatial outcomes and regional development on the one hand to the strategy and behaviour of business organizations on the other. This, in turn, set the stage for Dicken's focus during the 1980s on transnational business strategies and the global geographies of industrial change.

GLOBAL SHIFTS, TRANSNATIONAL CORPORATIONS AND INDUSTRIAL CHANGE

While maintaining his empirical research interests in the organization and impact of TNCs and the role of FDI in regional development (see Dicken, 1980; 1982; 1983; 1986b; 1987; 1988; Dicken and Lloyd, 1980), Dicken's most significant intervention of the 1980s was the publication of his first single-authored book *Global Shift* (Dicken, 1986a). Now in its fourth edition (Dicken, 2003), it has become one of the most widely used texts in the interdisciplinary study of global economic change (see Table 1.1). Comprehensive and ambitious, but at the same time eloquent, *Global Shift* can claim to be one of the pioneering globalization texts. Intentionally 'global' in perspective, the book aimed 'to describe and to explain the massive shifts which have been occurring in the world's manufacturing industry and to examine the impact of such large-scale changes on countries and localities across the globe' (Dicken, 1986a: vii). In this context, the main contributions of *Global Shift* can be traced in the following strands of literature: (1) spatial divisions of labour and global industrial change; (2) globalization studies; (3) international business studies.

Spatial divisions of labour and global industrial change

The period since the early 1970s had been one of turbulence in the international system. It was rapidly becoming clear that the global economy was undergoing an accelerated phase of restructuring, if not transformation, triggered amongst other things by the oil crisis in 1973 and the collapse of the Bretton Woods system of fixed exchange-rate regimes (see Tickell and Peck and Glasmeier and Conroy, in this volume). Manufacturing industries bore the brunt of these pressures. Indeed, such were the dramatic changes in the organization and geography of industrial activity that scholars in development studies began to talk in terms of a *new international division of labour* (NIDL) in which industrial production was seen to be shifting irreversibly from developed, industrialized economies to developing countries (Fröbel et al., 1980). *Global Shift* interrogated the central claims of the NIDL thesis, which at the time was on the way to becoming a new orthodoxy. Dicken's detailed analysis of the *range* of spatial and organizational shifts at work across different branches of manufacturing raised fundamental questions about the essential storyline of the NIDL thesis and its associated causative-cum-predictive claims. Through detailed case studies of four industries undergoing different kinds of global transformation (textiles and clothing, iron and steel, motor vehicles and electronics), Dicken (1986a) unpacked the over-generalized NIDL account, in part by pointing to the intervening effects of sector- and firm-specificity. Analytically, this provided the basis for a much more nuanced and multi-layered understanding of the processes of extra-national industrial change.

During the later part of the 1980s, another new literature was beginning to raise provocative questions about the process of global industrial change. This focused on the (allegedly systemic) transition from Fordism to post-Fordist flexible specialization (e.g. Piore and Sabel, 1984; Scott, 1988a). The post-Fordism debate foregrounded structural changes in the capitalist world economy, examining conditions surrounding episodic shifts in its dominant modes of accumulation and regulation. The movement towards post-Fordism was held to be leading to a series of profound changes in the technological, social, economic and territorial organization of production. Characterizing these in quasi-regulationist terms as 'technological-institutional systems', Scott and Storper (1992: 6) argued that their associated transformative processes were both historically and geographically specific. Although Piore and Sabel's (1984) seminal work *The Second Industrial Divide* was not cited in the first edition of *Global Shift*, Dicken was also thinking in terms of potentially transformative shifts in the macro structures of accumulation, though he continued to distance himself from more explicit post-Fordist arguments. *Global Shift* examined putatively systemic shifts in the context of a fourth Kondratiev wave during which 'really substantial *global* shifts in manufacturing production and trade have become apparent' (1986a: 22; original italics). Drawing upon the product life cycle theory, Dicken (1986a: 105) identified two major trends in the global production process in the 1980s:

1   an increasingly fine degree of specialization in many production processes, enabling their fragmentation into a number of individual operations;

2   the growing standardization and routinization of these individual opera-tions, permitting the use of semiskilled and unskilled labour.

Backed by detailed data on Ford and other automobile manufacturers, Dicken (1986a: 289–312) explored how flexible production systems had transformed the corporate strategies of automobile TNCs and the organization of the automobile industry itself: 'despite an undoubted degree of geographical dispersion of car and component manufacture by TNCs to developing countries, its extent and depth remain far more limited than might be expected from the nature of the motor vehicle industry itself' (1986a: 311). Extending his analysis to the electronics industry, Dicken (1986a: 336) observed 'the emergence of a remarkable geographical cluster of semiconductor and high-technology industries displaying all the characteristics of the classic Weberian locational agglomeration'. Of course, such arguments would later become central to emerging work on the locational logics of post-Fordism, debates around which were to play such an important role in the development of economic geography between the late 1980s and the mid-1990s (see Gertler, 1988; Tickell and Peck, 1992; Yeung, 1994; Scott, 2000).

Globalization studies

A second major strand of literature to which *Global Shift* has made a significant contribution might be broadly termed *globalization studies*. The concept of globalization has its obscure origins in French and American writings in the 1960s (see Held et al., 1999). Taylor et al. (2001) associated the ascendancy of 'globalization' as a millennial keyword with the formative statements of such business gurus as Levitt (1983) and Ohmae (1985; 1990). Although Taylor et al. (2001: 1) also noted that 'any intellectual engagement with social change in the twenty first century has to address this concept [globalization] seriously', geographers had rarely engaged with the globalization debate until the publication of *Global Shift* in 1986. While Dicken discussed the work of Levitt (1983), he also engaged with a range of earlier contributions on the internationalization of capital (e.g. Palloix, 1975; Harvey, 1982). Dicken's approach would characteristically focus on 'the emergence of a highly interconnected and interdependent *global* (rather than merely inter-national) economy', in the context of which he went on to explore the impacts and implications for 'nations and communities occupying different positions within this global structure' (1986a: 1–3; original italics).

Global Shift earned the double distinction of being one of the first books in economic geography to develop an explicitly *global* perspective and one of the first contributions to the nascent field of globalization studies to adopt a rigorously *geographical* orientation. For Dicken (1986a: 11; emphasis omitted), the book represented the first sustained attempt to explain 'the changing spatial form of [the changing] international political economy'. Prior to this, the majority of economic geographers had tended to focus on industrial change at the local and regional scales. As Leyshon (1994: 110) observed, by the end of the 1980s, ''the global' was firmly on the economic geography research agenda. The publication of the first edition of Dicken's *Global Shift* in 1986 was clearly an important milestone in this regard' (See also Dicken, Coe et al. and Tickell and Peck, in this volume.)

In the 1980s, the argument still needed to be advanced – and defended – that global level economic processes were more than the background scenery for national and local economic restructuring. By the 1990s, with the dramatic ascendancy of 'globalization speak' in political rhetoric as well as in academic research, it was necessary also to counter an argument at the other extreme, that globalization had somehow become an all-determining, omnipresent and unstoppable force. *Global Shift* was prey to neither of these fallacies, but instead carved out a more subtle and complex argument with respect to the nature, extent and consequences of global-level economic processes. In a sense, it established the case for the careful study of these 'global' processes in their own right, but it also advanced a deeper argument that national and local forms of economic restructuring were reciprocally related to globalization tendencies and that these must be conceptualized relationally (see also Brenner and Hudson, in this volume). *Global Shift*'s significant initial contribution was to open up the space for geographically orientated investigations of the global economic terrain and, thereafter, to establish some of the foundations upon which more nuanced, relational conceptions of economic transformation have been developed. The 1990s editions of the book were very much attuned to this analytical shift, helping to establish a persuasive counter-position to flat-earth/borderless world visions of globalization. In Dicken's view (also in this volume), nation-states remain key players in the structuring and regulation of the global economy; transnational corporations, more than simply bearers of a market logic, are engaged in dynamic, path-dependent and organizationally contingent processes of strategic development; global investment flows are not free-floating extra-terrestrial phenomena, but are embedded in networks and regulatory systems; globalization is a complex and uneven process, not a unilinear trend towards a unified international market. These arguments are sustained both analytically and empirically in *Global Shift*, which presents a reading of the globalization process that is quite distinct from the exaggerated accounts of the hyper-globalists on the one hand and the naysayers on the other (see Dicken et al., 1997a; Yeung, 1998a; 2002a).

The book was also being widely read outside geography, especially for its insistence that both spatial scale and uneven geographical development matter in attempts to grasp the complexity of globalization processes. Dicken's contention that globalization will, neither in principle nor in practice, lead to the erasure of national and local differences struck a chord – for all its planetary reach, globalization has myriad local and regional dimensions (see Dicken, 1986a: chapters 1 and 11–12). These insights have secured an influential role for *Global Shift* in the teaching of globalization in various disciplines, through its extensive adoption as a key text and within those policy circles that are concerned with promoting/managing globalization (e.g. the United Nations Conference on Trade and Development; see UNCTAD, 1994; 2001). It is also one of the few geographical studies cited in complementary works on globalization by other prominent social scientists (Dunning, 1993; Albrow, 1996; Hirst and Thompson, 1996; Held et al., 1999; Mittelman, 2000).

International business studies

A third major strand of literature in which *Global Shift* has cultivated a strong niche is *international business studies*. This literature differs significantly from the previous two fields primarily because it is much more applied in its research orientation. Whereas development studies, economic sociology and global political economy tend to dominate the first two strands of literature on globalization (see Table 1.1), international business studies draws on applied disciplines such as industrial economics, organization studies and strategic management. How then did *Global Shift* establish a presence in this field of applied business studies? First, while there was no shortage of geographical studies of TNCs completed in the 1980s (e.g. Taylor and Thrift, 1982; 1986), few of these studies had the analytical reach and empirical breadth present in *Global Shift*, which connected very suggestively with leading-edge work in international business studies, synthesizing insights from international business, including the product life cycle model and the eclectic framework of international production. Dicken proposed a conceptual framework for exploring alternative ways in which transnational production units could be organized. With his distinctive facility for graphical representation, Dicken conveyed a series of complex concepts relating to transnational business organization in a uniquely accessible form (see Figure 1.1), concepts that would be paralleled in Prahalad and Doz's (1987) integration-responsiveness framework, Bartlett and Ghoshal's (1989) 'transnational solution' and Dunning's (1993) conception of value-added TNC networks, while anticipating some of the recent theoretical work on organizational networks in the strategic management literature (Nohria and Ghoshal, 1997; Gnyawali and Madhavan, 2001).

Second, the explicitly geographical perspective in *Global Shift* enabled international business researchers to figure out the spatial implications of different corporate strategies and organizational structures pursued by TNCs. This said, the geography of international business activities has remained a rather serious research lacuna until recent years. As Dunning (1998: 46) has pointed out, '[t]he emphasis on the firm-specific determinants of international economic activity, while still driving much academic research by scholars in business schools, is now being complemented by a renewed interest in the spatial aspects of FDI'. The analysis of geography in international business studies, nevertheless, remains somewhat inadequate and under-developed, as it typically focuses on location and nationality of TNC activities (e.g. Kogut and Singh, 1988; Shaver, 1998; Nachum, 2000). An adequate analysis of TNCs and

Figure 1.1 Dicken on the geographical organization of TNC production units

a) Host-market production

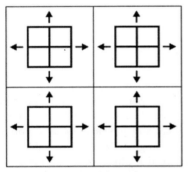

Each production unit produces a range of products and serves the national market in which it is located. No sales across national boundaries. Individual plant size limited by the size of the national market.

b) Product-specialization in a regional market

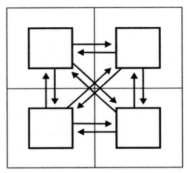

Each production unit produces only one product for sale throughout a regional market of several countries. Individual plant size very large because of scale economies offered by the large regional market.

c) Transnational vertical integration

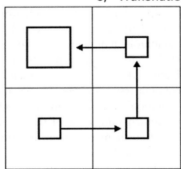

Each production unit performs a separate part of a production sequence. Units are linked across national boundaries in a 'chain-like' sequence – the output of one plant is the input of the next plant.

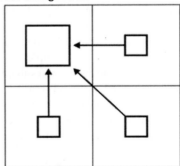

Each production unit performs a separate operation in a production process and ships its output to a final assembly plant in another country.

Source: Redrawn from Dicken (1986a: Figure 6.7; 203)

Figure 1.2 Dicken on the spatial evolution of TNC activities

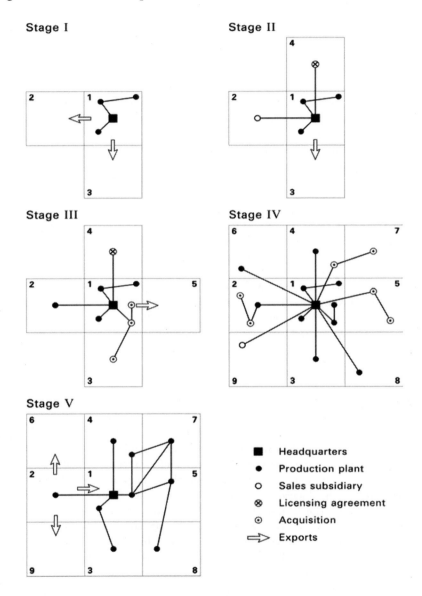

Source: Redrawn from Dicken (1986a: Figure 6.9)

FDI must incorporate both the locational shifts within TNC production networks (Dicken, 1986a: 212) *and* the geographically nested relationships from local to global scales through which these networks are constructed (Dicken, 1986a: 184). In an idealized model, for example, Dicken (1986a) (reproduced here as Figure

1.2), presented the spatial evolution of a TNC in relation to processes of reorganization, rationalization and spatial change.

Third, few books in international business studies offer detailed empirical analysis of TNC activities in such a comprehensive manner as *Global Shift*. Drawing upon massive amounts of carefully processed data and presenting these in a series of imaginative and informative figures and tables, the book came to exemplify, for an international business-studies audience, the best of geographical scholarship. Dicken's detailed analysis of how global shifts in a series of contrasting industries were shaped by the complex interactions between transnational production networks and nation-states led to favourable comparisons with the leading texts in international business studies (e.g. Dunning, 1993; Bartlett and Ghoshal, 1995; Punnett and Ricks, 1997; Stonehouse et al., 2000).

To sum up, while the book was written to describe and explain the turbulent industrial world of the 1980s, *Global Shift* was to make lasting contributions to the study of transnational corporations and global economic change. Dicken's work has clearly helped to create the interdisciplinary field of 'globalization studies'. This observation is confirmed by reviews of successive editions of *Global Shift*. In his review of the first edition, Krumme (1987: 132) postulated that 'this unconventional book might succeed in reestablishing some of the respect business schools used to have for geography courses'. By the third edition, Cox (1999: 475) verified that 'I know that Dicken's book, perhaps more than any other text, has done sterling service in bringing economic geography to the attention of students in other disciplines, particularly in international business courses.'

FIRMS, STATES AND GLOBAL NETWORKS

Subsequent editions of *Global Shift* were published in 1992, 1998 and 2003. In each successive edition, the reach of the book was extended. Empirically, its reach was extended beyond the manufacturing sector and into services. Analytically, there was a more explicit concern with 'the geographical unevenness of the economic, political and technological processes which, together, create global shifts in economic activity' (Dicken, 1998a: xiv). And in content as well as message, the book was becoming increasingly postdisciplinary in character. Pfirrmann (1999: 156), for example, has drawn attention to the way in which *Global Shift* engages with 'a broad set of interdisciplinary approaches and analytical paradigms to get into the complex causal relationships as well as the diverse tradeoffs concerning globalization's costs and benefits'. While the book's roots in economic geography remain evident, successive editions have repeatedly deepened the engagement with work in strategic management, global political economy and international business studies.

Parallel to the successive revisions of *Global Shift* during the 1990s, Dicken elaborated some of his early work on TNCs and nation-states as the principal actors in the global economy, venturing into the less familiar territory of *the political economy of firms and states* in the global space economy. Building on Stopford and Strange's (1991) characterization of firm-state relationships in terms of 'triangular diplomacy', Dicken argued that the tensions confronting TNCs and nation-states could be understood as differentiated power relations (see Dicken 1990a; 1990b; 1992b; 1994; 1995; 1997; 1998b). For TNCs, competitive pressures are generating dual tendencies, simultaneously to globalize operations in order to achieve greater efficiencies, while also localizing operations in order to ensure a degree of autonomy and responsiveness. For nation-states, conditions of accelerating globalization have been associated with far-reaching forms of institutional and functional reorganization, as '[t]he pressures towards certain kinds of putative supranational organization at one extreme are counterpoised against a pressure toward greater degrees of local political autonomy at the other' (Dicken, 1994: 122). The bargaining relationships between TNCs and nation-states are therefore very much situated in these complex global-local tensions. In a series of empirical papers published in the early 1990s, Dicken (1990c; 1992b; 1992c) showed how political and policy structures could promote and regulate the spatial strategies of TNCs and their FDI activities (see also Dicken and Tickell, 1992; Tickell and Dicken, 1993), with significant implications for uneven development and regional restructuring (Dicken and Quevit, 1994; Nilsson et al., 1996; Dicken et al., 1997b).

This focus on the continuing roles and the capacities of the nation-state amounts to an unambiguous refutation of 'the end of geography' thesis championed by such ultra-globalists as Ohmae (1990; 1995). In his contribution to *Twenty-First Century Economics*, Dicken (1998b: 41) argued that '[t]he nation-state remains a most significant force in shaping the world economy, for, although national boundaries may be far more permeable than in the past, as a territorial unit it continues to be the container of distinctive 'cultures' and institutional practices'. His argument for the continual importance of both geography in general and (reorganized) nation-state forms in particular has been echoed by other geographers (e.g. Cox, 1997; Yeung, 1998a; 2002a; Kelly, 1999; Amin, 2001). Relatedly, while many commentators in academia and policy circles have taken globalization to be a highly abstract and almost ethereal phenomenon, Dicken has steadfastly argued that globalization processes are initiated and mediated by economic and political *actors*. Ever since his 1971 article in *Economic Geography*, he has been concerned with firms (and later transnational corporations) as key drivers behind these ostensibly abstract processes. This firm-specific perspective on globalization is an important countervailing force in a field of study in which globalization is too often associated with under-specified and actor-less forces. In fact, much of the globalization literature has overlooked or over-simplified the issue of differentiated power relations between firms and states. By unpacking globalization processes, Dicken et al. (1997a; original emphasis) showed that globalization is 'articulated *through* both firms and states operating in complex interaction. Globalization does not exist as a free-floating structure, unrelated to the economic and institutional context in which it arises. It is constituted through those very practices which it subsequently transforms.' (See also Dicken, in this volume.)

In addition to these contributions to the globalization debate during the 1990s, Dicken worked with several contributors in this volume to conceptualize the *social organization* of business firms in terms of embedded networks and organizational relationships. His interest in these issues can be traced back to the notion of 'communications networks' in his 1971 article, in which he argued that '[t]he integration of the individual within these networks determines the amount and type of information to which he is exposed and helps to influence and modify the coding mechanism' (Dicken, 1971: 430). While there was some discussion of transnational corporate networks in the first edition of *Global Shift* (Dicken, 1986a: 203), the fullest articulation of his perspective on firms and networks was contained in a paper co-authored with Nigel Thrift (Dicken and Thrift, 1992). Here, concepts of embeddedness and networks were expounded for the first time in economic geography, partly in response to Walker's (1989) critique of corporate geography. These arguments helped pave the way to the so-called 'relational turn' in the subdiscipline, what is often styled in terms of the emergence of 'new economic geographies' (see Thrift and Olds, 1996; Lee and Wills, 1997; Clark et al., 2000; Sheppard and Barnes, 2000; Barnes, 2001; Yeung, 2003).

Drawing upon ideas from the 'new economic sociology' of Granovetter (1985), Zukin and DiMaggio (1990) and others, Dicken and Thrift (1992: 283) vigorously re-established the case for studying different organizational forms and processes in economic geography: 'the importance of organization as a cognitive, cultural, social and political (and spatial) framework for doing business has increasingly come to be realized. Indeed, nowadays, organization is often equated with 'culture', envisaged as a set of conventions.' In retrospect, this represented a telling move away from studying production *per se* towards a broader conceptualization of the socio-organization of production, prefiguring the extensive discussions that have taken place in recent years around 'network' paradigms, economies and geographies (see reviews in Yeung, 1994; 2000a).

Why exactly does embeddedness matter in economic geography? The answer, according to Dicken and Thrift (1992), lies with *networks* and their inherent *power relations*. They argued that understanding both production and capitalist social relations must start with embedded networks and relations in production systems because these 'processes do not simply occur in a general abstract form; they take on specific cognitive, cultural, social and political forms in an environment which is shaped very largely (although not exclusively) by business enterprises, especially large business enterprises which are able to wield more social power' (Dicken and Thrift, 1992: 284). This emphasis on networks and their associated power relations also has echoes in the recent 'rediscovery' of the firm in economic geography (see Taylor and Asheim, 2001), in part because it establishes an alternative analytical path between the methodological individualism of narrowly firm-centric approaches and the strong sense of structural predetermination that is evident in macro-process orientated studies of geographical industrialization. As Dicken and Thrift (1992: 286) put it, '[t]he analysis of intra- *and* inter-firm networks of power and influence is a very useful way of reconciling some of the differences between those who focus upon enterprises and those who, like Walker [1989], focus upon the organization of production in general'. This growing interest in embeddedness and networks in economic geography during the 1990s came to parallel ongoing contributions in economic sociology (see Velthuis, 1999) and in management and organization studies (Gulati and Gargiulo, 1999; Gnyawali and Madhavan, 2001).

GLOBAL PRODUCTION NETWORKS, TERRITORIAL ORGANIZATION AND RELATIONAL ANALYSIS

This most recent phase of Dicken's work has been associated with a renewed focus on theoretical development (see Table 1.1). One of his central concerns here has been the development of a *relational framework* for the study of global economic change, focusing on the character of inter-firm relations in different territorial and organizational contexts (see Dicken, 2000; Dicken and Malmberg, 2001; Dicken et al., 2001; Henderson et al., 2002). These relational networks are conceived both as *social structures* and as *ongoing processes*, which are constituted, transformed and reproduced through asymmetrical and evolving power relations by intentional social actors and their intermediaries. This relational view of networks emphasizes the role of human agency and the ongoing formation of networks that produce empirical outcomes and establishes a potential basis for what might be termed 'network ethics'. For example, thinking in terms of a global network for coffee or gold production or tourism services allows direct connections to be made between geographically distant consumers and producers, and the intermediaries in between (Whatmore and Thornes, 1997; Clancy, 1998; Hartwick, 1998; Olds and Yeung, 1999). In this way, the 'claims of distant strangers' (Corbridge, 1993) can become a part of economic and political analysis, rather than limiting such analysis to discrete political entities, like the nation-state.

By focusing on 'the essentially dialectical relationship between firms and places: the notion that *places produce firms* while *firms produce places*', Dicken (2000: 276; original emphasis) seeks in the context of firm-place nexus to operationalize some of the recent conceptual innovations concerning the territorial organization of the global economy that have been developed by geographers and sociologists (see Gertler, Malmberg, Brenner and Hudson, in this volume). Four deeply interconnected sets of firm-place relationships are identified in Dicken (2000: 285):

1   intra-firm relationships: between different parts of a transnational business network, as each part strives to maintain or to enhance its position vis-à-vis other parts of the organization;

2   inter-firm relationships: between firms belonging to separate, but overlapping, business networks as part of customer-supplier transactions and other inter-firm interactions;

3   firm-place relationships: as firms attempt to extract the maximum benefits from the com-munities in which they are embedded and as communities attempt to derive the maximum benefits from the firms' local operations and;

4   place-place relationships: between places, as each community attempts to capture and retain the investments (and especially the jobs) of the component parts of transnational corporations.

These ideas are further developed in collaboration with Anders Malmberg to incorporate transformations in the wider processes of territorial development at local, regional, national and global scales (Dicken and Malmberg, 2001). Here, the focus is placed on three major dimensions – firms, industrial systems and territories – embedded in the overall macro dimension of governance systems, which are characterized by prevailing sets of institutions, rules and conventions. Firms are conceived as complex spatial and territorial structures that simultaneously organize space into various localized industrial clusters while being influenced by the bounded nature of space.

REFLECTIONS

Peter Dicken has been a significant presence in economic geography for more than 30 years. During this time, this especially restless subdiscipline has gone through a series of theoretical and methodological shifts – from location theory through structural Marxism to a range of institutional, poststructural, cultural and relational turns (see Peck, 2000; Scott, 2000; Barnes, 2001; Yeung, 2003). In this rather turbulent intellectual context, Dicken's has been one of the more consistent and authoritative voices. Although his own position has evolved over time, his approach has remained 'grounded' – less concerned with theoretical fashions than with the challenging task of tracking and unpacking processes of global transformation in all their contingent complexity. And, as Ash Amin points out in this volume, Dicken has always done this with a 'light touch', managing to convey subtlety, intricacy and nuance, while avoiding the pitfalls of aimless empiricism, and charting global processes and transformations without succumbing to 'globaloney'. In this respect, as in others, he has cut his own path.

    Taking this broad view, there are a number of distinctive features in Dicken's work. First, he has always been concerned with *structures and processes* of global economic change. Right back to the first edition of *Global Shift* in 1986, he has made the case for a 'big picture' approach, with a geographical twist, to the study of global economic change. As pointed out by a reviewer of the second edition of *Global Shift*,

> … most of the big geographical questions associated with explaining the changing location of economic activity at the global scale were being more effectively addr-essed in the literature of political economy and business.

The publication of *Global Shift* in 1986 was therefore something of a landmark in returning these issues to their proper place within the geography curriculum. (Chapman, 1992: 134)

Second, while Dicken is always interested in the structures and processes of global shifts, he has never fallen into the trap of 'process fetishism', which can be defined as an exaggerated concern with processes at the expense of due concern with the actors and outcomes of globalization. In fact, his work has consistently shown how such actors as firms (particularly transnational corporations) and states matter in global shifts. This *actor-oriented approach* was a characteristic of his early work in the 1970s on the behaviour and decision-making processes of firms, while it also finds echoes in his more recent concern with firm-state relations and the role of lead firms in global production networks. Dicken's focus on the strategic behaviour of firms and states underscores the partiality of 'actor-less' interpretations of global economic change. Resistant to conceptualizations of the firm as an isolated 'island' of coordination within a sea of market relations – the classic Coasian transaction costs perspective, Dicken sees the firm as a kind of *relational* organizational device, connecting a structured field of actors and drawing upon a repertoire of resources to set and accomplish certain strategic initiatives. Cumulatively, his work represents a powerful analysis of the role of TNCs *qua* globalized firms, the locus of immense resources and boundary-spanning capabilities, as structuring agents in the globalization process (see also Gertler and Amin, in this volume).

Third, Dicken's distinctively *grounded approach* means that he is able to be empirical without being empiricist. Any user of *Global Shift* knows that this book is exemplary in its meticulous treatment of empirical materials. This affords the book a balanced and authoritative quality, irrespective of the reader's particular 'take' on globalization. For example, an online reviewer observed that *Global Shift*

> ... will challenge your view of globalization. Having come to the book as an opponent of globalization, this book awakened me to the complexity of the problems raised by a rapidly globalizing economy. As a result, I was forced to re-examine my opposition and hone my arguments against globalization. This unbiased, empirical approach makes the book highly recommended for those interested in putting forth the best possible arguments about the global economy.[2]

Similarly, an economist reviewing the third edition of *Global Shift* concluded that 'What I really like about Dicken is that he makes you think. He will take a topic like transnational corporations and look at all of their effects and the arguments and evidence before telling you what he thinks are the most important elements. He thus educates the reader to be a critical consumer of opinions (even his!) about globalization. He avoids the distorting generalizations that so often characterize globalization discussions.'[3]

Finally, one of the real virtues of Dicken's work is his *explicit interdisciplinarity*. Geography's relationship with other disciplines can often be somewhat asymmetrical, reflected in a reliance on 'imported' concepts, methods and frameworks (see Schoenberger, 2001), but Dicken has been amongst the 'exporters' too. His work not only echoes the concerns of disciplines like strategic management and global political economy, it also reciprocally speaks to these concerns. In a commentary on Dicken's (1994) Howard Roepke Lecture in Economic Geography reproduced in *Advances in Strategic Management*, Brahm (1994: 249 and 253) noted that the paper

> ... contains an insightful exploration of numerous topics regarding the global economy that are of great relevance to the field of strategic management ... I would like to praise Dicken's paper for its inter-disciplinary orientation and for its ambitious scope. It takes some courage to embark on this kind of endeavor knowing that an eclectic approach almost inevitably produces some fuzziness that discipline-specific purists will find dissatisfactory ... I hope that many strategy researchers will be influenced by it and will turn to some of the wider body of literature cited in the paper, including Dicken's other work, to enhance their own research agendas.

As a tribute to Dicken's formative contributions to studies of global economic change, we have briefly traced in this chapter some of the intellectual lineages of his work over the past three decades and pointed to some of its enduring influences. The contributors in this volume, in their own different ways, also draw attention to these diffuse and significant influences, tracking global connections in ways that are distinctively sensitive to space and scale. It is no exaggeration to say that Peter Dicken has been a pioneering presence in this field, and the range and depth of this work is in no small measure a tribute to his many contributions – personal and professional.

ACKNOWLEDGEMENT

With his characteristic modesty, Peter Dicken would probably have preferred that we did not write this chapter, so we would like to thank him for looking the other way long enough for us to get it done, and for his willingness to let us find our own way through his 'back catalogue'. We received helpful comments on an earlier draft of this chapter from Kris Olds, Nigel Thrift and Adam Tickell, but we must be solely responsible for any errors or misinterpretations that remain.

NOTES

1   *Location in Space*, co-authored with Peter Lloyd (Lloyd and Dicken, 1972; 1977; Dicken and Lloyd, 1990) remains as one of the standard teaching texts in economic geography. It has been translated into Italian (1979; 1993), Japanese (1997; 2001) and German (2001). Intriguingly, it was one of the very few economic geography

works cited in Porter's (1990) study of *The Competitive Advantage of Nations*. For his significant contributions to advancing research on globalization and economic geography, Peter Dicken was awarded the Victoria Medal by the Royal Geographical Society-Institute of British Geographers in 2001. The citation reads 'Peter Dicken's scholarly work both encompasses the globe and is celebrated around it by leading academics and institutions both within and beyond Geography as agenda-setting and leading research which is innovative, critical and sustained' (*Geographical Journal*, Vol.167, Part 3, September 2001, pp.271-272). In 2002, he was awarded an honorary doctorate of philosophy by the University of Uppsala, Sweden, in recognition of his standing as one of the world's most distinguished economic geographers.

2   Tim Hundsdorfer, http://www.amazon.com, accessed on 12 December 2000.
3   Michael Veseth, http://www.ups.edu/faculty/veseth/reading.htm, accessed on 24 December 2001.

PART ONE

GROUNDING GLOBAL FLOWS

Chapter 2

'PLACING' FIRMS:
GROUNDING THE DEBATE ON THE 'GLOBAL' CORPORATION

Peter Dicken

INTRODUCTION

The purpose of this chapter is to engage with *one* aspect of the globalization debate: the relationships between firms (specifically transnational corporations) and geographical space. Most writers on globalization project a highly simplistic conceptualization of the firm that spans the ideological spectrum: from the 'hyper-globalists' of the more populist business literature to those of the anti-globalization movements. All engage in an unhelpful form of reductionism by seeing transnational corporations (TNCs) as becoming essentially homogeneous entities that are, in effect, *placeless*: the so-called *global corporation*. My argument is that place and geography still matter fundamentally in the ways firms are produced and, therefore, in how they behave. This, I believe, is more than merely an academic issue; it is also one with potentially significant policy implications in the context of economic development at different geographical scales. The chapter is organized into three major parts together with a conclusion. The first part sets out briefly some of the assertions on global corporations from both the pro- and anti-globalization camps. The second part examines some attempts to measure the 'globalness' of TNCs. The third section addresses the ways in which firms are 'placed', particularly in terms of the geographical context within which they originate.

GLOBAL CORPORATIONS RULE, OK?

The view from business

One of the central claims of the 'hyper-globalists' in business is that international firms are inexorably, and inevitably, abandoning their ties to their country of origin and, by implication, converging towards a universal *global* organizational form. Kenichi Ohmae's (1990: 94) infamous exhortation to business managers is usually invoked as the exemplar of such a position:

> Country of origin does not matter. Location of headquarters does not matter. The products for which you are responsible and the company you serve have become denationalised.

Ohmae's position reflects a pervasive view among many writers today. But it is not a new idea. The US Under-Secretary of State in the 1960s, George Ball, coined the label 'Cosmocorp' to denote what he saw as the then emerging global corporation (Ball, 1967). Barnet and Muller (1974) provided numerous anecdotal examples of the intentions of US corporate executives to transform their firms into 'placeless' global corporations. More recent musings by busi-ness leaders and commentators convey rather vague, sometimes quite bizarre, ideas as to what a 'global' corporation actually is (or should be). Airport book-stalls are full of such 'literature'. Their underlying argument is, essentially, that technological and regulatory developments in the world economy have created a 'global surface' on which a dominant organizational form will develop and inexorably wipe out less efficient competitors who are no longer protected by national or local barriers. Such an organization is, it is argued, 'placeless' and 'boundary-less'.

The claim that the placeless corporation is, or is becoming, the norm amongst international business firms received a substantial boost in the 1990s with the persistence of the Japanese financial crisis, following the collapse of the bubble economy at the beginning of the decade, and the equally unexpected East Asian financial crisis of 1997–1998. In effect, the model of the global corporation became, to all intents and purposes, that of the *US corporation*. US market capitalism had seemingly triumphed; the neoliberal IMF/Washington/Wall Street consensus sets the rules (see Tickell and Peck, in this volume). It now seems a very long time ago that the US business and political communities were obsessed with the apparent failure of the US corporate model in the face of the Japanese variety of capitalism (and, indeed, of the newer forms emanating from Korea, Taiwan, and other parts of East Asia). Shareholder value became the hot ticket. For some, indeed, it became the only show in town. The US-style corporation was projected as being the most effective way of maximizing such value. All other models of business organization were not only less efficient (some were even corrupt), but would, inevitably, be vanquished in a neo-Darwinian struggle (see also Gertler, in this volume).

However, the events of 2002 – notably the ignominious collapse of Enron, WorldCom, and other high-profile US companies – seriously threw into doubt both the efficiency and incorruptibility of the US corporate model. Indeed, these events provide an ironic commentary on the 'holier-than-thou' stance of many US business writers who attributed a major role to 'crony capitalism' in the East Asian crisis.

More generally, it does seem to be the case that most of the world's largest business corporations *think* of themselves as either being 'global' or, at least, 'globalizing'. In his analysis of the 'transnational capitalist class', Sklair (2001) interviewed senior figures in around 80 of the 500 largest corporations in the world in order to try to establish their use of, and the meaning they attached to, the term 'globalization'.

The suggestion that multinationals were 'national companies with units abroad' was roundly rejected as old fashioned and not compatible with the demands of the contemporary global economy. Most Global 500 executives and managers in the sample considered their corporations to be in a transitional state between the multinational corporation and the global corporation, that is, they were to a greater or lesser extent globalizing. Also clear was the finding that, in order to fulfil a 'shareholder-driven growth imperative', most of the corp-orations considered that they had no alternative but to globalize... All of these findings demonstrate a move to globalization among the Global 500 and a certain level of success. Nevertheless, few of these corporations considered themselves entirely globalized (Sklair, 2001: 73).

The views of the anti-globalizers

The notion that the world is now dominated by massive global corporations is not solely confined to the world of business. Many writers on the political left tend to take a similar view (although, of course, with a very different interpret-ation of the economic, social, and political implications). One of the oldest devices is to compare the size of TNCs with nation-states in order to demons-trate that TNCs have become more powerful than states (see, for example, Benson and Lloyd, 1983: 77). Its most recent manifestation is a survey published by the Institute for Policy Studies in the US (Anderson and Cavanagh, 2000: 3), which makes the following claims, amongst others:

* 'Of the 100 largest economies in the world, 51 are corporations; only 49 are countries (based on a comparison of corporate sales and country GDPs)... To put this in perspective, General Motors is now bigger than Denmark; DaimlerChrysler is bigger than Poland; Royal Dutch/Shell is bigger than Pakistan.'

* 'The 1999 sales of each of the top five corporations (General Motors, Wal-Mart, Exxon Mobil, Ford Motor, and DaimlerChrysler) are bigger than the GDP's of 182 countries'.

These are, of course, very striking comparisons. But it is far from clear what they really mean. In particular, as Wolf (2002: 19) has pointed out, the comparison is based on totally different criteria:

> GDP is a measure of value-added, not sales. If one were to compute total sales in a country one would end up with a number far bigger than GDP. One would also be double-, triple-, or quadruple counting ... if corp-

orations, too, are measured by value added as national economies are...they tend to shrink by between 70 and 80 per cent. In 2000, sales by General Motors were $185bn but value added was $42bn; sales by Ford were $170bn but value added was $47bn; and sales by Royal Dutch/Shell were $149bn but value added was only $36bn... Properly measured, Denmark's economy is more than three times bigger than GM. Even impoverished Bangladesh has a bigger economy than GM.

So, in evaluating the scale of global corporations, such comparisons are not very helpful. They do, of course, perform the extremely valuable function of forcing us to think about the shifting relative power between private and public institutions with all its political and economic implications – that, after all, is the purpose of the Institute for Policy Studies. Such comparisons, therefore, have real value as polemic, but they do not tell us much about the 'global-ness' of corporations, or even the extent to which corporations are more, or less, oriented to domestic or foreign operations.

MEASURING 'GLOBAL-NESS'

All of these different views beg the question of how we might define a 'global' corporation – a far from simple task. In my view, a global corporation would be

> a firm that has the power to co-ordinate and control operations in a large number of countries (even if it does not own them), but whose *geographically-dispersed operations are functionally-integrated,* and not merely a diverse portfolio of activities.

Unfortunately, this is a rather demanding definition in empirical terms. So, let's see what we can do with the available evidence (see also Gertler, in this volume).

The most comprehensive published data on TNCs at the global scale are those published annually in UNCTAD's *World Investment Report.* Since its 1993 Report, UNCTAD has developed a composite *Transnationality Index* (TNI). This is simply a weighted average of three indicators: foreign sales as a percentage of total sales; foreign assets as a percentage of total assets; and foreign employment as a percentage of total employment. These data have now been published for seven successive years, from 1993 to 1999, allowing some comparison of trends over time. Analysis of these data reveals a number of important features:

1   The mean TNI for all 100 TNCs in 1999 was 52.6. In 1993, the mean was 51.6. In other words, for the top 100 TNCs as a group – albeit, of course, not a completely identical group because of entries and exits over the intervening period – there was no significant increase in their degree of transnationality. Hence, we cannot say that this elite group – what we could reasonably regard

as the world's global corporations – have, by this simple measure at least, become more global.

2   The fact that the TNI index for this group of firms is only a little over 50 suggests that, on average, these firms have roughly half their operations in their countries of origin and half abroad. This does not suggest an especially high degree of globalization.

3   Only 57 of the 100 companies in 1999 had a TNI of greater than 50 and only 16 companies had a TNI greater than 75 (i.e. indicating more than three-quarters of their activities outside their countries of origin).

4   There is no correlation between the size of a firm's foreign assets and its TNI (see Figure 2.1).

5   Only two companies – Nestlé and the Nippon Mitsubishi Oil Company – appear in the top 15 on the basis of both foreign assets and their TNI. In general, the largest TNCs in terms of total foreign assets all have relatively low TNIs (see Table 2.1). For example, as the largest TNC in terms of assets, General Electric ranks 75th on the TNI list. GM, 4th in terms of foreign assets, is 83rd on the TNI ranking. For Ford, the numbers are 5th and 77th for Toyota, 6th and 82nd and for IBM 9th and 50th

6   There are some substantial differences in the degree of transnationality between firms from different home countries. Not surprisingly, in general, firms from smaller countries (either geographically or in terms of their economic size) tend to have higher TNIs than firms from larger countries (see Table 2.1).

7   Figure 2.2 shows the pattern of TNIs for the leading home countries for both 1993 and 1999. A number of features are evident:

•   Although the mean TNI for US firms increased from 36.7 in 1993 to 41.5 in 1999, the overall degree of transnationality of US firms is lower than all the other major home countries other than Japan. In 1999, only one-third of the 25 US firms in the list had a TNI greater than 50, with the highest value being 66.

•   Similarly, Japanese firms have a low degree of transnationality on average. The mean value increased from 33 in 1993 to 38.4 in 1999. Again, around one-third have a TNI less than 50.

•   Amongst European firms, those from the UK have, on average, the highest TNIs. In 1993, UK TNCs' mean TNI was 68.6; by 1999 it had risen to 75.5.

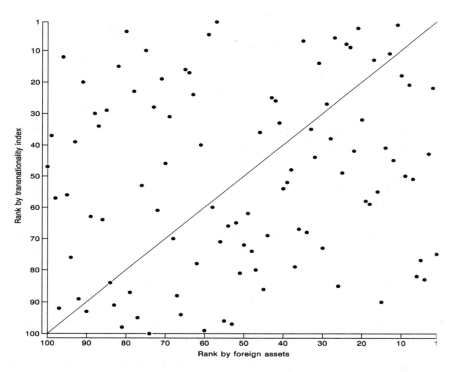

Figure 2.1 The relationship between firms ranked by transnationality index and foreign assets, 199

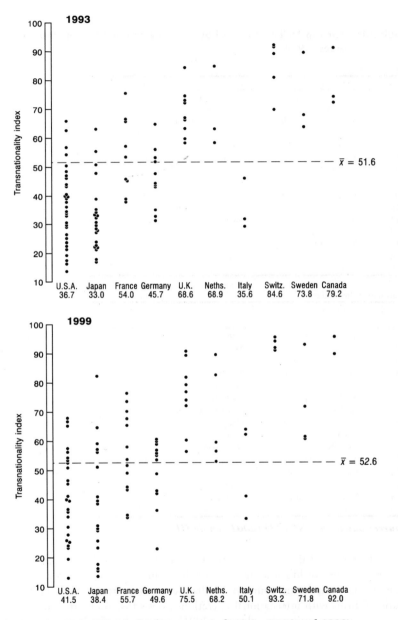

Figure 2.2 ransnationality index by country of origin, 1993 and 1999

Source: Calculated from UNCTAD (1993; 2001).

Table 2.1 **The top 15 TNCs ranked by foreign assets and transnationality index, 1999**

Foreign Assets

Rank	Company	Country	TNI rank
1	General Electric	United States	75
2	ExxonMobil	United States	22
3	Royal Dutch/Shell	United Kingdom	43
4	General Motors	United States	83
5	Ford	United States	77
6	Toyota	Japan	82
7	DaimlerChrysler	Germany	51
8	TotalFina	France	21
9	IBM	United States	50
10	BP	United Kingdom	18
11	Nestlé	Switzerland	2
12	Volkswagen	Germany	45
13	Nippon Oil Co	Japan	11
14	Siemens	Germany	41
15	Wal-Mart	United States	90

Transnationality Index

Rank	Company	Country	Foreign assets rank
1	Thomson Corporation	Canada	57
2	Nestlé	Switzerland	11
3	ABB	Switzerland	21
4	Electrolux	Sweden	80
5	Holcim	Switzerland	59
6	Roche Group	Switzerland	27
7	BAT	United Kingdom	35
8	Unilever	UK/Netherlands	24
9	Seagram	Canada	23
10	Akzo Nobel	Netherlands	75
11	Nippon Oil Co	Japan	13
12	Cadbury-Schweppes	United Kingdom	96
13	Diageo	United Kingdom	17
14	News Corporation	Australia	31
15	L'Air Liquide Grp	France	82

Source: Based on UNCTAD (2001: Table III.1).

The results of this analysis of the UNCTAD data are clear in one respect: the majority of the world's leading TNCs still retain more than 50 per cent of their activities in their home country. In that sense they are, in Hu's (1992) terms, 'national firms with international operations', notwithstanding the views expressed by corporate executives in Sklair's (2001) survey described earlier. The major exceptions are firms from small home countries. But even if the figures did show a greater degree of 'transnationality', they would not tell us very much because the measure used is simply a dichotomous one: home versus abroad. It can tell us nothing about the real geographical extent of a firm's activities. At least in principle, a firm could have a TNI of, say, 80 (meaning that 80 per cent of its activities were outside the firm's home country), but all of those activities could be in one foreign country. An example would be the significant number of US firms that operate in Canada but not elsewhere. So, in fact, all the TNI and similar measures can do is to measure the extent to which a firm's activities are domestic or foreign-based.

However, the UNCTAD data do throw light on some specific issues, particularly some of the conclusions of Anderson and Cavanagh's (2000) comparison of TNCs and nation-states referred to earlier. In particular, the Anderson and Cavanagh list, based on *sales*, contains a much larger proportion of US firms (41 per cent) than does the UNCTAD list based upon *transnationality* (25 per cent). In other words, United States firms are subst-antially less global than Anderson and Cavanagh's study implies. Hence, the UNCTAD appears to be a much better measure of the global-ness of firms. One specific example can be used to illustrate this discrepancy. According to Anderson and Cavanagh, Wal-Mart is the biggest private employer in the world with 1,140,000 workers. The implication is that this has a major impact globally. However, if we look more closely at Wal-Mart, we find that the extent of its transnational operations is very limited indeed. Although Wal-Mart is the largest retailer in the world in terms of sales, its sales outside the US represent only 14 per cent of its total sales. Likewise, Wal-Mart may well employ over one million workers, but around 85 per cent of these are located in the US. Wal-Mart's operations outside the Americas are very limited indeed. In Europe, they are confined to the UK and Germany; in Asia to a small number of stores in China and Taiwan. It is hard, then, from a non-US perspective, to regard Wal-Mart as a 'global' corporation.

However, there is a slight flaw in this argument. The TNI (and similar data) are based on *ownership* criteria. They provide a measure of the geographical extent of a TNC's directly owned assets, but they tell us nothing about the *functional* organization of their activities – not just those directly owned, but also those coordinated and controlled, even where not owned. So, for example, Wal-Mart may have a relatively limited global spread in terms of its stores, but still be a more influential global player because of its *global sourcing policies*. Exactly the same would apply to many, though not all, the leading TNCs (see also Schoenberger, in this volume). In other words, the UNCTAD data provide only a small window on these firms' operations.

So, all the TNI (and similar measures) do is to measure the extent to which a firm's directly owned activities are domestic or foreign-based. They tell us nothing about the geographical spread of these operations or about the indirect influence of such firms. We can only really explore these through company-specific investigations. As far as geographical spread is concerned, examination of firms' annual reports or websites tends to show that only a relatively small number of firms have globally spread operations. Rather more tend to have strong *regional* biases in their operations (see Hudson, in this volume).

'PLACING' FIRMS

The quantitative data discussed in the previous section are useful for my present purposes because they demonstrate the continuing real ties between firms and their home countries. The world's leading TNCs remain strongly enmeshed in their home country environments. But the data do not tell us anything about the nature of those ties: about the *qualitative* nature of TNC activities and their relationship to place. Neither do they help us to establish whether or not TNCs of different national origins are becoming similar in their modes of operation. It is at least possible that TNCs may retain more of their assets and employment in their home country, but still be converging organizationally and behaviourally towards a universal, global form. Equally, some firms may be becoming more global, but in non-uniform ways. If so, what role does the firm's country of origin play in this?

The response of geographers to assertions of the rise of the placeless global corporation has been to re-assert – somewhat reflexively, even defensively – the importance of place. But we haven't done very much more either to think through what this involves conceptually or to provide robust empirical evidence. The theoretical basis for hypothesizing that TNCs 'produced' in different places will continue to display a significant degree of organizational differentiation lies in the much used (and sometimes abused) concept of *social embeddedness* (Dicken and Thrift, 1992; Dicken et al., 1994).

Although such embeddedness may occur at a variety of geographical scales, the most significant scale would appear to be that of the *national state*, the major 'container' within which distinctive practices develop. The term 'container' should not be taken too literally. It is used here as a fairly loose metaphor to capture the idea that nation-states are one of the major ways in which distinctive institutions and practices are 'bundled together'. Of course, such containers are not (except in very rare cases) hermetically sealed off from the outside world. The container is permeable or leaky to varying degrees. It can be argued that the impact of the modern information technology and communic-ations systems has made national containers even more permeable. That's true – but it does not mean that the container no longer exists at all. Indeed, there is a good deal of compelling evidence to show the persistence of national distincti-veness – although not necessarily uniqueness – in structures and practices which help to shape local, national and international patterns of economic activity.

In addition to 'culture' (see, for example, Hofstede, 1980; 1983), three characteristics are especially important in defining the context within which firms develop (Doremus et al., 1998):

1   the dominant ideology of the state;

2   the nature of its political institutions; and

3   the nature of its economic institutions.

Taken together, these dimensions help to define what Whitley (1992: 13) calls distinctive *business systems*: 'distinctive configurations of hierarchy-market relations which become institutionalized as relatively successful ways of organizing economic activities in different institutional environments'.

Such 'national containers' of distinctive assemblages of institutions and practices help to 'produce' particular kinds of firms. In the TNC literature, Dunning (1979) was one of the first to make the explicit connection between what he terms the 'ownership-specific' advantages of firms and the 'location-specific' characteristics of national states. However, some thirteen years earlier, Vernon's (1966) model of the product life cycle had been based on the implied relationship between the nature of the US domestic economy and the foreign investment behaviour of US firms. Like all writers in the FDI theoretical literature, both Dunning and Vernon owed an immense debt to the pioneering work of Stephen Hymer (1960/1976). Dunning's (1979: 284) view, however, was that 'as an enterprise increases its degree of multinationality, the country specific characteristics of the home country become less, and that of other countries more, important in influencing its ownership advantages' (see also Nachum, 2000 for related ideas). I will return to this point.

Influence of home country environments

Despite these general observations on the hypothesized relationships between firms and the national contexts in which they are embedded, there have been very few systematic comparative empirical studies. Those that do exist give some support for the idea that firms from different countries differ, at least in part, because of variations in their national institutional structures. For example, Sally (1994) finds significant differences between French and German TNCs. Biggart and Hamilton (1992) and Yeung (1998b) critically analyse the differences between Asian and Western businesses. Whitley (1999) explores the 'divergent capitalisms' of East Asian economies. Gerlach and Lincoln (1992) compare Japanese and American business networks. I will use three recent cross-country comparative studies to illustrate this argument in more detail.

The first is the substantial study entitled *The Myth of the Global Corporation* by Doremus et al. (1998) (see also Pauly and Reich, 1997). These writers explicitly confront the 'convergence' thesis through a detailed empirical comparison of American, German, and Japanese TNCs along a series of structural and behavioural dimensions, including their modes of corporate governance, corporate financing systems, and their strategic behaviour (notably in relation to R&D, direct investment, and intra-firm trade). In the case of American, German and Japanese firms, their conclusions are unequivocal:

> [There appears to be] little blurring or convergence at the cores of firms based in Germany, Japan, or the United States ... Durable national institutions and distinctive ideological traditions still seem to shape and channel crucial corporate decisions ... the domestic structures within which a firm initially develops leave a permanent imprint on its strategic behavior ... At a time when many observers emphasize the importance of cross-border strategic alliances, regional business networks, and stock offerings on foreign exchanges – all suggestive of a blurring of corporate nationalities –

> our findings underline, for example, the durability of German financial
> control systems, the historical drive behind Japanese technology
> development through tight corporate networks, and the very different time
> horizons that lie behind American, German, and Japanese corporate
> planning. (Pauly and Reich, 1997: 1, 4, 5, 24)

As far as German firms are concerned, these conclusions are supported by a study
of German companies operating in Britain and Spain (Ferner et al., 2001). They
found the persistence of distinctive German practices in human resource
management and industrial relations (see also Gertler, in this volume).

My second empirical example is drawn from Yeung's (2002b) recent
comparative analysis of firms from Hong Kong and Singapore. Yeung shows how
the different political and institutional histories and environments of the two city-
states have produced distinctive entrepreneurial characteristics, even though, in
both cases, their ethnicity is comparable (mainly Chinese).

> In Hong Kong, the Chinese business system is the dominant mode of
> business operations. The spirit and ethos of capitalism in Hong Kong have
> produced socially, culturally, and politically specific business systems ...
> Ethnic Chinese industrialists in Hong Kong are known for their entre-
> preneurship and higher propensity to engage in risky business and overseas
> ventures ... The peculiar neoliberal political economy in Hong Kong had
> several consequences for transnational entrepreneurship in Hong Kong.
> First, the private sector assumed a leading role in Hong Kong's economic
> development ... Second, the lack of direct state intervention in Hong
> Kong's industrialization and economic development processes has con-
> tributed to the growth of domestic companies in both large firm sectors and
> small firm networks ... Third, the financial system in Hong Kong ... is
> highly favourable for the development of the service sector. (Yeung, 2002b:
> 98)

In contrast, although the Chinese mode of business is clearly present in Singapore,
it differs in some important respects from the situation in Hong Kong. The main
reason, in Yeung's (2002b: 99) view, is the very different institutional structure in
Singapore.

> Notably, a large proportion of local investments, particularly in the
> manufacturing sector, came from foreign firms and GLCs [government-
> linked companies] ... and their various subsidiaries. The role of indige-nous
> private enterprises in Singapore's industrialization is rather limited ... The
> majority of Singaporeans have become contented with their job security and
> are less willing to take specific kinds of risks to launch new business
> ventures.

My third empirical example is drawn from the comparative study of South
Korean and Taiwanese firms carried out by Hamilton and Feenstra (1998). There
are very strong similarities in the history and developmental experiences of South
Korea and Taiwan. Their business firms also share common features of Confucian-

based familism: 'In all these background variables – economic, social, and cultural – Taiwan and South Korea are as nearly the same as could be imagined between any two countries in the world today. Yet the economies of these two countries are organized in radically different ways' (Hamilton and Feenstra, 1998: 124). Although business organizations in both economies are organized as networks of family-owned firms, their modes of organization differ considerably.

In South Korea, the dominant type of business group is the *chaebol* (conglomerates). The model for such firms was the Japanese *zaibatsu*, the giant family-owned firms which had been so important in the pre-Second World War development of the Japanese economy. According to Wade (1990: 124), the *chaebol* 'are highly centralized, most being owned and controlled by the founding patriarch and his heirs through a central holding company. A single person in a single position at the top exercises authority through all the firms in the group. Different groups tend to specialize in a vertically integrated set of economic activities.' As a result, the South Korean economy became highly concentrated and oligopolistic, while the small- and medium-sized firm sector is relatively underdeveloped. Not only this, but many of these smaller firms are tightly tied into the production networks of the *chaebol*. Indeed, the *chaebol* have developed as some of the most highly vertically-integrated business networks in the world: 'the firms in the *chaebol* are the principal upstream suppliers for the big downstream *chaebol* assembly firms ... in Samsung Elect-ronics, most of the main component parts for the consumer electronics division are manufactured and assembled in the same compound by Samsung firms' (Hamilton and Feenstra, 1998: 128–29). Taiwanese business networks, in contrast, have low levels of vertical integration. The more horizontal Taiwanese networks consist of two main types: 'family enterprise' networks and 'satellite assembly' networks (independently owned firms that come together to manu-facture specific products primarily for export).

These contrasts between the South Korean and Taiwanese business groups – despite the strong similarities between the two economies – have been explained as arising from

> differences in social structures growing out of the transmission and control of family property. In South Korea, the kinship system supports a clearly demarcated, hierarchically ranked class structure in which core segments of lineages acquire elite rankings and privileges. These are the 'great families' ... In Taiwan, however, the Confucian family was situated in a very different social order... Unlike Korea (and in the early Chinese dynasties), where the eldest son inherited the lion's share of the estate and all the lineage's communal holdings, in late imperial China the Chinese practiced partible inheritance, in which all sons equally split the father's estate ... This set of practices preserved the household and made it the key unit of action, rather than the lineage itself ... In summary, although based on similar kinship principles, the Korean and the Chinese kinship systems operate in very different ways. (Hamilton and Feenstra, 1998: 134–35)

Such differences in socio-cultural practices largely explain the contrasts between the ways that business firms are organized in the two neighbouring economies.

Cross-country comparisons are essentially aggregative in scale. A rather different way of looking at the problem is to examine the extent to which differences in firm behaviour exist within specific industries. I refer briefly to two recent pieces of research. The first is the electronics industry in East Asia. A recent UNCTAD (2002: 103) study found

> significant differences in the way Japanese and United States firms organize their production networks in East Asia, particularly with regard to the location of management, the sourcing of components and capital goods, the replication of production networks, and the motive for investing abroad.

US electronics firms tended to use more market-based management relationships with relatively more independent affiliates and to employ more local personnel in higher-level technical tasks than Japanese firms, which relied more on intra-firm arrangements within *keiretsu* and tended to use more expatriate Japanese personnel in senior positions.

A second example, also from East Asia, is drawn from research into the Indonesian garment industry (Dicken and Hassler, 2000). Significant differences exist in the manner in which firms from different countries of origin operate within the Indonesian garment industry. Firms from East Asian countries (notably Taiwan, Hong Kong and South Korea) tend to establish direct manufacturing operations in Indonesia. In contrast, American and European firms tend to operate through networks of local agents and traders. There are further observable differences between American and European firms in the extent to which they establish their own representative offices or use independent agents:

> Fieldwork evidence suggests that, in general, European buyers prefer to work with their own representative offices based in Jakarta. These representative offices are endowed with decision making power, and prefer to procure materials domestically because of the possibility of internalising the means of control at earlier stages in the production chain … In contrast, most United States companies (with the exception of the very large firms, such as Levi and Nike) have shown a preference for working with East Asian agents rather than establishing their own representative offices … Agents have representative offices in most of the countries involved in garment manufacturing; the head office remains in the major cities of the 'first generation' NIEs such as Seoul, Hong Kong, or Taipei. The US buyers deal mainly with the head office and transfer all responsibilities to the agent, including quality control. (Dicken and Hassler, 2000: 272)

Influence of host country environments

These cross-country and intra-sectoral comparisons demonstrate that significant differences in firm behaviour do exist and that these can be related to the national context in which such firms originate. However, we should not assume that, because the conditions in which firms develop in their home country environments exert an extremely powerful influence on their behaviour, the impact of the *host environments* in which they operate is no longer important (see Dicken, 2000; Dicken and Malmberg, 2001; Schoenberger, 1999). Indeed, for a whole variety of reasons – political, cultural, social – non-local firms invariably have to adapt some of their domestic practices to local conditions. It is virtually impossible, for example, to transfer the whole package of firm advantages and practices to a different national environment with different institutional and regulatory practices. Abo (1994), for example, points to the 'hybrid' nature of Japanese overseas manufacturing plants. The same argument applies to US firms operating abroad. Even in the UK, where the apparent 'cultural distance' between the US and the UK is less than in many other cases, there is a very long history of American firms having to adapt some of their business practices to local conditions. Wever (1995) analyses the industrial relations difficulties faced by German firms operating in the US and of US firms operating in Germany. Gertler's (2001; also in this volume) analysis of the experiences of German machinery manufacturers in Canada similarly shows the importance of different macro-regulatory and institutional contexts for the transposition of German production systems.

Whether or not, as Dunning (1979) claimed, increasing degrees of transnationality reduce the influence of home country context on firm behaviour as a whole, is debatable. Much will depend on the extent to which the experiences of individual affiliates of TNCs, located in different parts of the world, are incorporated into the TNC's overall modes of operation, that is, how far intra-organizational learning actually occurs (see Gertler and Amin, in this volume). There is little evidence to suggest that this creates *fundamental* changes in firms' overall behaviour. Although TNCs undoubtedly have to make adjustments to specific local conditions, they appear to do so to a relatively marginal degree.

CONCLUSION

The central argument of this chapter has been that, contrary to much received wisdom, place and geography still matter fundamentally in the ways in which firms are produced and in how they behave. My basic point is that firms – including TNCs – are 'produced' through an intricate process of embedding in which the cognitive, cultural, social, political, and economic characteristics of the national home base play a dominant part. TNCs, therefore, are 'bearers' of such characteristics, which then interact with the place-specific characteristics of the countries and communities in which they operate to produce a set of distinc-tive outcomes. The Russian painter Marc Chagall once observed that 'every painter is born somewhere, and even if later he responds to other surroundings, a certain essence, a certain aroma of his native land will always remain in his work' (Notes to an Exhibition at the Royal Academy of Arts, London, on 'Chagall: Love and the Stage', 1998). It seems to me that Chagall's observation is a better metaphor of the relationship between TNCs and place than Ohmae's (1990) vision of the placeless corporation. It more sensitively captures the complexity of the embeddedness process in which both place of origin and the other places in which TNCs operate influence the ways in which such firms behave and how they, in turn, impact upon such places. Within this essentially dialectical relationship, however, the TNC's *place of origin* appears to remain the dominant influence.

This is not to claim that TNCs from a particular national origin are identical. This is self-evidently not the case; within any national situation there will be distinctive corporate cultures, arising from the firm's own specific corporate history, which predispose it to behave strategically in particular ways. Neither does this imply that nationally embedded business organizations are unchanging. On the contrary, the very inter-connectedness of the contemporary global economy means that influences are rapidly transmitted across boundaries. This will, inevitably, affect the way business organizations are configured and behave. There 'is essentially a process of co-evolution through which different business systems may converge in certain dimensions and diverge in other attributes' (Yeung, 2000b: 425).

For example, the *keiretsu* have been at the centre of Japanese economic development during the post-war period. But the financial crisis in Japan that has persisted since the bursting of the 'bubble economy' at the end of the 1980s has put them under considerable pressure to change at least some of their practices. In particular, the recent influx of foreign capital to acquire significant, sometimes controlling, shares in some of these companies has had a catalytic effect. The most notable example was the acquisition by the French automobile company, Renault, of 36.8 per cent of the equity of Nissan. There are strong pressures, particularly from western (notably US) finance capital, for the Japanese business groups to open up to outsiders, to reduce or eliminate the intricate cross-shareholding arrangements, and to become more like western (i.e. US) firms with their emphasis on 'shareholder value' rather than the broader socially based 'stakeholder' interests intrinsic to Japanese companies. While, without doubt, some changes are occurring, it would be a mistake to assume that Japanese firms will suddenly be transformed into US-clones. The Japanese have a very long history of adapting to external influences by building structures and practices that remain distinctively Japanese.

Similarly, South Korean and other East Asian firms have come under enormous pressure to change some of their business practices in the aftermath of the region's financial crisis of the late 1990s. In South Korea, the *chaebol* are being drastically restructured and the relationships with the state reduced. Among Chinese businesses, the strong basis in family ownership and control is being challenged both by internal and external forces. Greater involvement in the global economy is forcing these firms to modify some of their practices (see Yeung, 2000b: 411–24). And yet it would be extremely surprising if the distinctive nature of nationally based TNCs were to be replaced by a standard-ized, homogeneous form despite the dreams of the hyper-globalists.

Hence, despite the unquestioned geographical transformations of the world economy, driven at least in part by the expansionary activities of transnational corporations, we are not witnessing the convergence of business-organizational forms towards a single 'placeless' type.

> Surface similarities in the behavior of ... [TNCs] ... abound ... [but] At root ... the most strategically significant operations of ... [TNCs] cont-inue to vary systematically along national lines. The global corporation, adrift from its national political moorings and roaming an increasingly borderless world market is a myth ... The empirical evidence ... suggests that distinctive national histories have left legacies that continue to affect the behavior of leading [TNCs] ... The scope for corporate inter-dependencies across national markets has unquestionably expanded in recent decades. But history and culture continue to shape both the internal structures of [TNCs] ... and the core strategies articulated through them. (Doremus et al., 1998: 3, 9)

This is because, over time, and under specific circumstances, societies have tended to develop distinctive ways of organizing their economies, even within the broad, apparently unitary, ideology of capitalism. Not all capitalisms are the same; capitalism, like the famous brand of canned food, comes in many different varieties (see, for example, Berger and Dore, 1996; Hollingsworth and Boyer, 1997; Whitley, 1999). Not only this, but such distinctive forms tend to persist over time, even though they may become modified 'at the margin' through interaction with other social systems of production:

> forms of economic coordination and governance *cannot* easily be transferred from one society to another, for they are embedded in social systems of production distinctive to their particular society ... Economic performance is shaped by the entire social system of production in which firms are embedded and not simply by specific principles of management styles and work practices ... institutions are embedded in a culture in which their logic is symbolically grounded, organizationally structured, technically and materially constrained, politically defended, and historic-ally shaped by specific rules and norms.

There are inherent obstacles to convergence among social systems of production of different societies, for where a system is at any one point in time is influenced by its initial state. Systems having quite different initial states are unlikely to converge with one another's institutional practices. Existing institutional arrangements block certain institutional innovations and facilitate others:

> Despite the emphasis on the logic of institutional continuity, this is not an argument that systems change along some predetermined path. There are critical turning points in the history of highly industrialized societies, but the choices are limited by the existing institutional terrain. Being path dependent, a social system of production continues along a particular logic until or unless a fundamental societal crisis intervenes. (Hollingsworth, 1997: 266–68)

Organizational diversity, related at least in part to the place-specific contexts in which firms evolve, continues to be the norm. That, at least, seems to me to be a reasonable interpretation of existing empirical evidence. But that should not be seen as the end of the story. Much of the evidence we have is suggestive rather than conclusive. There is a real need to unravel the complexities of firm-place relationships in a more theoretically sophisticated and empirically rigorous manner. Such a research agenda needs to involve more than an intellectual exercise *per se*. If, as I have argued in this chapter, firms from different geographical contexts are different in significant respects then this has enormous implications for economic development policy at national, regional and local levels. To understand these implications requires meticulous comparative international analyses of firm-place relationships. In this regard, such research forms part of the broader intellectual effort devoted to understanding the nature and characteristics of the persistently varied and divergent forms of capitalism. However, the conclusions based on the evidence so far suggest that country of ownership continues to matter a lot for the behaviour of TNCs. They remain, to a very high degree, products of the local 'ecosystem' in which they were originally planted. TNCs are not placeless; 'global' corporations are, indeed, a myth.

ACKNOWLEDGEMENT

This chapter draws upon and develops some of the arguments in Dicken (2000). I am grateful to participants at seminars and workshops in Boulder, Leipzig, Liverpool, and Uppsala for their extremely helpful comments. I want to thank Jamie Peck and Henry Yeung both for their comments on this chapter, but – most of all – for producing this book. I shall be eternally grateful.

Chapter 3

GLOBALIZATION, TRANSNATIONALISM AND THE ASIA-PACIFIC

Neil M. Coe, Philip F. Kelly and Kris Olds

INTRODUCTION

Few geographers have done more than Peter Dicken to identify, synthesize and analyse the dizzying complexity of contemporary change in the global economy. While transnational corporations have been the primary locus of Peter's work, he has consistently addressed their roles and strategies with a geographical sensibility. This has meant recognizing the distinctiveness of place and region, the social embeddedness of capitalist logic(s), the importance of state and non-state institutions, and the impacts of corporate strategy and capitalist restructuring for the people whose lives are involved. Peter Dicken's work has therefore embraced the firm, the network, the state, the region, the worker and the manager as its multiple scales of analysis.

An important part of this project has been the development of an understanding of contemporary globalization: as political and cultural discourse; as overstated conceptual construct; as empirical phenomenon and as corporate strategy. In each case, Peter has provided a clear-headed analysis of the process that serves as a corrective to both economistic generalizations and politically charged representations (Dicken, 1998a; Dicken et al., 1997a).

Over the last ten years, however, a new stream of scholarship has emerged that also seeks to address the deepening integration of societies and their economies. Its provenance has not been in economics, management or political science – the cognate disciplines for much of Peter's work - but in sociology, anthropology and the humanities. It concerns the transnationalism created by large (but not unprecedented) numbers of people moving across national borders over the last 30 years, creating in the process 'transnational social fields'. Thus, when sociologists, anthropologists and social geographers now talk of transnationalism, they are as likely to be talking of migration flows of various kinds and the transborder social spaces thereby established, as about the border-crossing trade, investment and production networks of corporations.

This chapter seeks to integrate the insights available from this recent literature with those from Peter Dicken's *oeuvre* on the globalization of capital, trade and production. We argue that keeping the two apart occludes many realities in the contemporary global economy, where the movement of people, and their social networks across space, are both shaped by, and constitutive of, global economic restructuring. Such a conceptualization can productively explore the complex interactions between the various structures and agents that shape the global economic system.

The chapter is structured in the following way. In the next section, we define contemporary transnationalism, identify some of its key dimensions and show how it has been largely separated from analyses of the global economy. The third section traces the connections between migrant transnationalism and global economic restructuring. Four distinct linkages are identified. The final section provides two empirical case studies from the Asia-Pacific region to illustrate some of these linkages in the context of two distinctive spatial forms in the changing global economy – world cities and new industrial spaces. The first case study highlights the importance of social networks in the channelling of global capital flows – in this case, property investment between 'world cities' on opposite sides of the Pacific. The second case study draws attention to the formation of new transnational communities within the information technology industries of the Asia-Pacific.

DEFINING TRANSNATIONALISM

The conceptual vocabulary now used to analyse the global economy is often confused and usually overlapping: globalization, internationalization, multinational, transnational. The notion of *transnationalism* has also, however, been adopted in relation to phenomena comprising the linkages maintained between places by migrants:

> We define transnationalism as the process by which immigrants forge and sustain multi-stranded social relations that link together their societies of origin and settlement. An essential element is the multiplicity of involvements that transmigrants sustain in both home and host societies. (Basch et al., 1994: 6)

The novel element in the transnationalism literature since Basch et al.'s seminal book has been to rethink the conventional understanding of migrant workers and immigrants. Until recently, the usual treatment of 'people flows' was as either contract workers purely responsive to the demands of global capitalist growth, or as immigrants who broke their ties with their homelands and settled with various degrees of success in new homes. The late 1990s saw, however, a flourishing field that identified the emergence of transnational social spaces created by international migration and facilitated by intensifying transportation and communications linkages. Immigrants of all classes were recognized as having transnational lives – economically, culturally, socially and politically (see, for example, Hannerz, 1996; Portes et al., 1999; Portes, 2001; Levitt, 2001; Smith, 2001).

As with globalization, many have argued that such transnationalism is far from new and indeed barely warrants a new conceptual vocabulary. There are, however, distinctive elements of recent transnationalism that deserve particular attention. First, technologies such as air travel, global mail and courier services, international financial transactions, telephones, fax machines and emails have allowed a speed and intensity of interaction across global space that is unprecedented. Second, evidence suggests that there are increasing numbers of immigrants, and the non-migrants they leave behind, who are closely tied to transnational networks for both their cultural identities and economic livelihood. As Portes et al. (2001: 4) note, 'ceasing to be exceptional, transnational act-ivities may become common and even normative'. Third, transnational linkages between migrant populations and their 'homes' are receiving increasing attention from a growing number of states, from El Salvador to the Philippines. Administrative structures to facilitate labour export, remittance flows and return migration have become important components of some national development strategies (Gonzales, 1998), while the attraction of talented migrants has been a mainstay of other development strategies – for example, Singapore's 'foreign talent' programme or Canada's Business Immigrant Scheme (Coe and Kelly, 2000; Ley, forthcoming).

Despite the growing importance of transnational ties predicated largely on migration, the two dimensions of globalization have usually been treated as separate processes – global capitalist logic, corporate strategy and state or supra-state regulation as one set of concerns, and migration, financial remittances, cultural hybridity and immigrant settlement another. Seldom have connections been made between the two. Indeed it is a commonplace in the literature on transnationalism to encounter a problematic binary distinction between 'transnationalism from above' (referring to institutionalized and corporate actors) and 'from below' (meaning migrants, contract workers, expatriates, households etc.) (Smith and Guarnizo, 1998).

Sassen's (1996) work on labour, capital and urbanization processes stands out as an attempt to overcome this dichotomy. By focusing on the urban dimensions of both global capital and global migration, Sassen explores some of the interconnections between the two. Global cities, she argues,

> ... are the sites for the overvalorization of corporate capital and the further devalorization of disadvantaged economic actors, both firms and workers. The leading sectors of corporate capital are now global in their organization and operations. And many of the disadvantaged workers in global cities are women, immigrants, and people of color, whose political sense of self and

> identities are not necessarily embedded in the 'nation' or the 'national community'. Both find in the global city a strategic site for their economic and political operations. (Sassen, 1996: 206–7)

Sassen suggests that viewing international capital and labour (i.e. immigration) in the same geographical frame in this way highlights the connections between the two. The internationalization of labour is, she argues, the counterpart of the internationalization of capital, and the activities and spaces at the vanguard of the global economy in terms of their attraction to capital are also fundamentally predicated upon an internationalized labour market (Sassen, 1988).

MIGRANT TRANSNATIONALISM AND THE GLOBAL ECONOMY: TRACING THE CONNECTIONS

With a few exceptions, the studies of transnationalism noted above have largely taken the individual, household or community as their locus of analysis, and have therefore been divorced from the literature on global capitalist restructuring. Our purpose here is to make this connection more explicit by arguing that the transnational flows and linkages noted above can be linked far more closely with global shifts in production, investment and trade. In this way, we seek to apply Dicken et al.'s (2001) network epistemology and his concern with structure and agency in the global economy. There are at least four ways in which these connections might be made.

Remittances, migration and corporate transnationalism

At the household level, studies have demonstrated the effects that hard currency remittances can have upon class, status and gender relations within and among households (for example, Conway and Cohen, 1998). At the macro-economic scale, meanwhile, these personal remittances amount to a fundamental component of some national economies. The structural economic significance of migration and remittances goes further, however, than these direct economic implications. In a study of Jamaican data entry operators working in transnational branch plants or subcontractors, for example, Mullings (1999) shows how household finances are subsidized by dollar remittances from North America with two effects. First, they act as a supplement to the very low wages paid by employers in the country's export processing zones, thus implicitly subsidizing such corporate operations by financing the reproduction of a workforce being paid less than a social wage. Second, such remittances permit a level of resistance to workplace discipline that would be hard to sustain without a safety net to depend upon in times of unemployment. The result has been a socio-spatial stereotyping of the Jamaican workforce as 'difficult' by corporate locators. In both ways, then, the household forms an important nexus connecting transnational capital and production with migration and remittances.

Social networks and investment flows

It is increasingly recognized that investment flows do not simply follow an inexorable economic logic, but instead are guided by and travel along social networks. Here, work on overseas Chinese (or ethnic Chinese) business networks has been especially significant in highlighting the importance of personal ties for economic integration. Taiwanese and Hong Kong investment in the People's Republic of China and in Southeast Asia, for example, is directed by, and dependent for success upon, the social and cultural ties linking investors and local partners (Hsing, 1998; Smart and Smart, 1998; Yeung, 1998c; Hamilton, 2000). Further afield, investment from Hong Kong and Singapore into Vancouver and other Pacific Rim cities has been motivated by, and channelled through, family and social networks (Mitchell, 1995; Olds and Yeung, 1999; Mitchell and Olds, 2000; Olds, 2001). These networks create the conditions for acting at a distance with some degree of reliability, especially in cross-cultural contexts.

Global restructuring and (im)migrant workers

Much of the literature on global economic restructuring has pointed to the emergence of new industrial spaces, world cities and regional economies as nodes of trade and investment in the global economy (Scott, 2001; see Malmberg, in this volume). Too often, however, the dependence of such places on other kinds of global flows – in particular, people – is neglected. This is so at many different levels of class and status. As an aspirant global city, for example, Singapore is highly dependent on two streams of migrant workers: a low-paid and low-status army of workers imported temporarily to provide labour forces in construction, childcare/domestic work and manufacturing; and a high-paid, high-status contingent of expatriate professionals in the finance, high-technology and creative industries (Coe and Kelly, 2000; Hui 1997; Yeoh and Chang, 2001).

Equally, as Sassen (1988) has shown, the global cities of the core – the control and command centres of the global economy – are dependent on an underclass of service workers, mainly immigrants, to build, clean and serve them. Indeed as Portes (2001) and others have noted, the logic of capitalist restructuring in the 'developed world' constantly demands low cost and marginalized immigrant labour as a structural response to competition and a 'spatial fix' of sorts against crises of accumulation. At the opposite end of the wage and status hierarchy, the immigration of skilled technicians has been a key element in the development of many new industrial spaces – the role of Asian migrants in Silicon Valley will be described later in the second of our case studies below. Furthermore, these immigrant-dependent labour markets feed the remittance economies described above. It is therefore possible to imagine a scenario in which a janitor in the head office of a major transnational corporation in New York (or London, or Sydney) sends money back to his wife (or sibling or parents) in the Philippines (or Mexico, or Pakistan). This money subsidizes the wages she receives at the firm's local production facility, which in turn pay for another relative to raise the capital needed to finance their own immigration application, and so the process continues.

A final transnational community associated with economic restructuring comprises a cosmopolitan class of elite professionals (see, for example, Beaverstock, 1996). Coming from a world-systems standpoint, Sklair (2001) has described what he calls a transnational capitalist class, comprising four key components: the executives of transnational corporations and their local affiliates; globalizing state bureaucrats; globalizing politicians and prof-essionals; and consumerist elites. The common feature uniting these groups is that they operate internationally as a normal part of their working lives. There is also considerable mobility between the groups. Together, they constitute a global power elite bound together by the culture-ideology of consumerism. While quite different from the immigrant workforce of a global city, these actors similarly live and work transnationally. It is this transnational social formation that is examined in the first of our case studies below.

Migration and transnational entrepreneurship

As a consequence of both the transnational social networks created by migration and the marginalization of immigrants in host societies (and particularly urban labour markets), entrepreneurship is often stimulated. Landolt et al. (1999) identify four types of enterprises that might result from such transnational connections:

1   'Circuit' firms provide services transporting or transmitting goods and remittances (or in facilitating the migration process itself, for example as recruitment agencies).

2   'Cultural enterprises' import and sell cultural goods (magazines, music, videos etc.) from the 'home country'.

3   'Ethnic enterprises' supply goods such as food and clothing from 'home' to both immigrant and non-immigrant urban markets.

4   'Return migrant microenterprises' are established by returnees who utilize contacts made while overseas.

The possibilities within these four categories are extensive – 'ethnic' clothing for sale in cosmopolitan downtown boutiques and 'exotic' foods in grocery stores are two obvious examples. But transnational entrepreneurship may also extend to high-tech start-ups utilizing connections in India to recruit programmers or subcontract production processes; or the marketing of real estate transnationally (for example, subdivisions in the Dominican Republic sold to immigrant communities in New York, or, condominiums in Vancouver sold exclusively in

Hong Kong) (see Itzigohn et al., 1999; Olds, 2001). Such activities are not corporate transnationalism in the conventional sense, but represent the integration of transnational social and cultural ties with a form of capitalist enterprise.

A great deal more could be said about each of these issues and other linkages might be identified, such as the influence of 'transnational advocacy networks' upon firms and institutions regulating the global economy (Evans, 2000) or the creation of demand for consumer products due to social and cultural remittances from migrants to their home communities and the related creation of yearnings to migrate by the consumption aspirations introduced by transnational corporations (Appadurai, 1996). Nevertheless, these notes point clearly enough to our broader argument, which is that analyses based on the institutional, macro-economic and corporate dimensions of restructuring are closely, and increasingly, linked to the emergence of a transnational social, cultural, economic and political space based upon mass migration. They imply that the separation of globalization and transnationalism, 'from above' and 'from below', institutional and informal, is often inappropriate.

TWO CASE STUDIES

In the remainder of this chapter, we will illustrate this argument through two case studies. The first highlights the importance of social networks, transnationalized in the process of migration, in directing flows of capital investment, in this case in Pacific Rim urban property markets. The second draws attention to the development of new transnational communities connecting Silicon Valley with emerging technology regions in Asia through the migration and circulation of highly skilled technical and engineering staff.

Transnationalism and property development processes in Vancouver and Toronto

Of all the various 'capitalisms' that exist in Pacific Asia, it is 'Chinese capitalism' that is the most deterritorialized: stretched, as it were, throughout Pacific Asia (Hamilton, 1996; 1999; Dicken and Yeung, 1999) and increasingly across the Pacific into cosmopolitan cities like Vancouver, Toronto and San Francisco. Immigration plays a key role in extending the social relations that constitute Chinese capitalism, whether the extensification process is permanent, temporary or episodic in nature. In diverse and complex ways, the operation and spread of Chinese capitalism to countries like Canada and to cities like Vancouver and Toronto is carried by the migration of entrepreneurs; entrepreneurs who are part of households; households that, collectively, give Chinese capitalism its characteristics. This is because the principal 'carriers' of Chinese capitalism have been, and continue to be, *households* via family-run enterprises/firms.[1]

Given the above points, Figure 3.1 highlights the disaggregated individuals, some of whom (for not all Hong Kong immigrants are entrepreneurs) comprise the extended households that have effectively stretched Chinese capitalism across trans-Pacific space in the late twentieth century.

Figure 3.1 Hong Kong immigration flows to Canada, 1962–2000

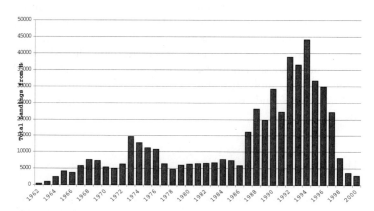

Sources: Olds (2001: 78) and Citizenship and Immigration Canada.

Immigration plays a fundamental role in enabling entrepreneurs to move across space, enabling them to become embedded in new (additional) locales, and to engage in economic activity (Ley, forthcoming). Were it not for selective immigration policies in Canada (and the US), the transnationalism that binds together multiple territories on the Pacific Rim would not have facilitated the economic activity that it has. For good and for bad, transnationalism underlies and shapes global economic change.

In this evolving process, segments of economic sectors in Vancouver and Toronto have become incorporated into the evolving form of social and economic organization known as Chinese capitalism. In legally migrating from Hong Kong to Canada, the most economically active of these extended households, elite capitalist families from Hong Kong, have generated powerful transnational social fields that have shaped economic (and especially urban) development processes in Vancouver and Toronto since the late 1980s.

As ever, contingent historical events must also be factored in. The structural context of monopolistic conditions and hyper-profits in Hong Kong's property market was jarred by the convergence of two East Asian geopolitical events – the 1984 Sino-British accord regarding Hong Kong's future and the 1989 Tiananmen crisis – that heightened insecurity. Simultaneously, for different reasons, Canada's immigration system was reformed. This reform, described in Mitchell (1995) and Olds (2001), opened Canada to a wider range of people from countries such as Hong Kong and Taiwan, especially if they possessed relatively large sums of capital or educational and professional qualifications. The parallel movements of geopolitics in East Asia and immigration policy reform in Canada therefore enabled (and inspired) many Hong Kong entrepreneurs or their extended family members to acquire Canadian citizenship, especially during the 1980s and 1990s (Skeldon, 1994).[2]

More specifically, large-scale urban redevelopment projects (see Table 3.1 below) were initiated in Vancouver and Toronto in the late 1980s and mid-1990s, all of which are associated with extended (trans-Pacific) families and associates linked to elite Hong Kong-based Chinese business firms. These projects have also provided impetus for the acquisition of older and more established Canadian firms, including some of the largest listed property firms in the country. In just over a decade, then, transnationalism has facilitated the reworking of downtown residential landscapes in two large cities and also reshaped the corporate hierarchy within Canada's property development sector.

Table 3.1   Canadian urban development projects associated with trans-Pacific families

Project	City Place, Toronto	Coal Harbour, Vancouver	Pacific Place, Vancouver
Size	18 hectares	41 hectares	80 hectares
Cost ($CDN)	$1.5 billion	$2 billion	$3 billion
No. of residential units	7000	2200	8500
Development timeline	1997–2005/2010	1988–2005/2010	1988-2005/2010
Developer(s)	Concord Adex Development Corp.	Aspac Developments Ltd.; Marathon Developments; Delta Land	Concord Pacific
Main Hong-Kong/Canadian families involved	Li family; Hui family	Kwok family	Li family; Hui family
Main Hong Kong firms associated with these families	Cheung Kong; Hutchison Whampoa; Wing Hong Contractors	Sun Hung Kai	Cheung Kong; Hutchison Whampoa; Wing Hong Contractors

For example, the most well known of these projects – Pacific Place in Vancouver – was acquired by the Li family of Hong Kong. This project, tightly intertwined with the professional education of the eldest son (Victor Li) of Li Ka-shing, is now 60 per cent built (Olds, 2001). Victor Li acquired Canadian citizenship in 1983 at the age of 18. Victor, as well as numerous employees and supporters of the firm, symbolizes the hybrid nature of the transnational, trans-Pacific elite, with educational qualifications acquired on both sides of the Pacific, several homes on both sides of the Pacific, political connections and support on both sides of the Pacific and stellar linguistic capacities. Take, for example, the initial director and senior vice-president of Concord Pacific, Stanley Kwok. Kwok, a Chinese Canadian who migrated to the country in 1968, was the perfect intermediary in a process involving global and local processes, publics and people. The mesh between the global and the local is dependent upon cultural hybrids like Kwok who use their expert knowledge to interpret local conditions and negotiate cultural differences for the geographically distant capitalists and financiers (Mitchell, 1997).

Pacific Place preceded a series of subsequent investments in Vancouver by prominent Hong Kong families who also acquired their wealth via the operation of Hong Kong-based property conglomerates. As noted in Table 3.1, Coal Harbour Project is a CDN$2 million project. This project was initiated by Marathon Realty of Canadian Pacific Railways, one of Canada's oldest companies. At the height of Hong Kong–Canada migration flows in the early 1990s, Marathon entered into a joint venture with Aspac Developments to develop the 41-hectare waterfront mega-project. Aspac Developments is a private firm established by the Kwok family, the main shareholders of the Sun Hung Kai conglomerate in Hong Kong. While the majority of the Kwok fortune is invested in Asia and managed out of Hong Kong, they have a series of smaller development projects in Canada as well as numerous familial linkages across the Pacific, some of which were established decades ago when Canada's immigration system experienced its first major 'opening' to the Pacific (in the late 1960s). For example, one close relative of the Kwoks, Thomas Fung, migrated to Canada in 1967 during the first 'wave' of immigration across the Pacific (see Figure 3.1). Fung has since become one of Canada's 'media barons' with a media and property development empire (the Fairchild Group) worth approximately CDN$190 million. The Fairchild Group is the developer of a series of Asian malls in the Vancouver suburb of Richmond and manages Canada's only national Chinese language media network. Fung's father, Fung King Hey, co-founded the sprawling Sun Hung Kai empire in the early 1970s via Sun Hung Kai Securities and Sun Hung Kai Bank (Dolphin, 1995; Thomas, 1995).

Trans-Pacific socio-economic networks are ever evolving and constantly reconstituted. For example, the drivers of the Pacific Place project (the Li family) have withdrawn back to Hong Kong because of the significance of Hong Kong and China to their core listed firms (Cheung Kong and Hutchison Whampoa). In other words, the trans-Pacific transnationalism associated with the eldest of the Li sons (Victor Li) has been affirmatively withdrawn, replaced by a regional (Pacific Asian versus Pacific Rim) and global (finance) transnationalism because of the demands associated with operating Hong Kong- and New York-listed conglomerates that are heavily involved in China, a series of Southeast Asian countries, as well as global ports (via Hutchison International Terminals) around the world.

On linked fronts, however, the Pacific Place development (and the development firm, Concord Pacific Developments) has been used by families (especially the Hui family) and firms associated with the Li family to further deepen linkages with property markets in Vancouver, Toronto and elsewhere in Canada. This expansion process has involved the integration of Hong Kong-based Chinese capital with 'Canadian' capital via acquisitions and mergers, all planned by firms with exceedingly strong access to knowledge resources in the financial arena and capital resources that can be leveraged upon in Hong Kong. A series of complex corporate manoeuvres were implemented between 1992 and the present that led a network of Chinese families to deepen their integration into the Canadian property market sector. A shell company was acquired in 1992, renamed to Burcon International Developments in 1993 and then used to conduct a reverse takeover of Regent Park Realty (a Vancouver real estate brokerage). Burcon was then used as a Vancouver-based platform to integrate into larger scale Hong Kong capital (most notably the Li family's Hutchison Whampoa). Burcon was capitalized via Bermudian and Hong Kong initiatives and then used to take over a failing Canadian property company, Oxford Properties Canada. This 1995 transaction was recorded under the Investment Canada Act [*http://icnet.ic.gc.ca/investcan/jun95.htm*], an act designed 'to provide for the review of significant investments in Canada by non-Canadians in order to ensure such benefit to Canada', implying the admission of 'business immigrants' to the country. Increasingly complex and labyrinthine manoeuvres were utilized to enhance the financial strength of the network of firms associated with Burcon and Oxford Properties. One key manoeuvre involved the acquisition of Burcon shares by Concord Pacific in 1996, the initiation of the City Place development project in Toronto (see Table 3.1) by Concord in 1997, the listing of Concord Pacific in 1998 on the Vancouver Stock Exchange, further mergers between networks of linked firms and, most recently (in 2001), the sale of the main profit-generating entity (Oxford) to the Ontario Municipal Employees Retirement System (OMERS) for large scale returns.

This series of complex financial and organizational movements demonstrates the knowledge and capital resources available to the trans-Pacific entrepreneurs who were allowed into Canada in three main waves – the late 1960s, the mid-1970s and the late 1980s/early 1990s (again, see Figure 3.1). The process of firm establishment and subsequent mergers and acquisitions has been underlain in a variety of ways by complex and deep linkages via immigration from Hong Kong to Canada. Key entrepreneurs for all of the firms noted in this section have acquired Canadian citizenship and/or have extended family and friendship ties with Canada (mainly anchored in Toronto and Vancouver). Many of these entrepreneurs have also retained significant linkages in Hong Kong, for Hong Kong's role as a global city (Sassen, 1998), and as place of economic calculation, is pre-eminent, especially compared to any Canadian city. In short, transnationalism is intertwined with the extension of economic practices and with the reworking of the habitus of global business.

Transnationalism in Pacific Rim information technology industries

Our second case study concerns the burgeoning transnational communities that are an integral part of the information technology (IT) industries of the Pacific Rim. While there is a wide range of partial and anecdotal information on the topic, Saxenian (1999) provides the first systematic and detailed exposition on the scale and significance of these communities. In particular, she demonstrates the increasingly significant role that skilled immigrants are playing in the economic development of Silicon Valley: currently over one third of research scientists and engineers in Silicon Valley's high-technology workforce are foreign born, with the majority being of Asian descent. By 1998, immigrant-run technology companies (started since 1980) accounted for more than US$16.8 billion in sales and 58,000 jobs. There were 2,001 firms led by an ethnic Chinese CEO and 774 with an ethnic Indian CEO, together accounting for 24 per cent of the total number of firms, The rate of immigrant new firm formation has increased steadily since the early 1980s. The continued success and dynamism of Silicon Valley is thus increasingly predicated on trans-Pacific flows of highly skilled migrants – particularly from Taiwan, India and China, but also in much lesser numbers from Vietnam, the Philippines, Japan and Korea – who contribute both directly to the economy, as engineers and entrepreneurs and indirectly as traders and middlemen linking California to the technologically advanced areas of Asia.

Nation-states play a key role in constructing and shaping these flows. The vast majority of these high-skill migrants have arrived in California since 1965, when the Hart-Cellar Act replaced the highly restrictive 'nation of origin' quotas for immigration with a system that allowed immigration based on scarce skills (and on family ties to existing residents). Immigration rates rose in line with the growth in demand for skilled labour from the region's electronic and computer industries. The Immigration and Nationality Act of 1990 further increased the number of visas offered on the basis of occupational skills. The debate about how many 'H-1B' visas – for foreign workers in specialty occupations that require a college degree – are to be offered annually is still hugely significant. Three main flows over the last 30 years or so are worth noting: the Chinese engineering workforce is dominated by an influx of Taiwanese in the 1970s and 1980s; immigration from mainland China took off in the 1980s and accelerated in the 1990s; and immigration from India has become the single most important element since the 1980s. A considerable proportion of these migrants head for California and, in particular, Silicon Valley. Others come to California to study for undergraduate or graduate degrees and then take up jobs in the region's technology companies. These 'new immigrant entrepreneurs' (Saxenian, 1999) have responded to the challenges of their new environment by setting up their own companies in increasing numbers and organizing collectively.

The success of Asian migrants in Silicon Valley is thus partly founded upon their local networks. These migrants are very different to the historical image of immigrants as lone pioneers and in reality rely upon a diverse range of informal social structures and institutions. While those who arrived during the 1970s and early 1980s undoubtedly saw themselves as outsiders, over time local social and professional networks have emerged that enable the mobilization of the know-how and capital needed to start high-tech firms. Saxenian (1999), for instance, identifies two Indian and eleven Chinese professional organizations established in Silicon Valley since 1980. It appears that the most successful immigrant entrepreneurs have both leveraged these ethnic connections *and* integrated themselves into local technology and business networks. Concomitantly, however, there has been a 'globalization' of Silicon Valley's ethnic networks, a phenomenon that is of particular interest to us here. In effect, many of these new firms, although small, are 'global' actors from day one, with the ability to use ethnic network connections to access Asian sources of capital, development skills and markets (Shin, 2001). It is these linkages that take the story beyond one of simple brain-drain migration and cheap offshore production to one that encompasses the formation of transnational technical communities characterized by *mutually beneficial* connections and the *circulation* of people, capital, technologies and ideas. These links are especially apparent between the technology communities of the Valley and Taiwan, and through Indians who are becoming key middlemen in linking Californian companies to low-cost software expertise in India.

Hsu and Saxenian (2000) provide a fascinating case study of the developing links between Silicon Valley and Taiwan's predominant technology region, Hsinchu. They argue that the growth of ties between firms and actors in Hsinchu and Silicon Valley is part of the competitive advantage of *both* regions. The key actors are a transnational community of US-educated Taiwanese engineers who have both the experience and language skills to operate effectively in both regions. The dense social and professional networks of these engineers enable two-way flows of technology, capital, know-how and information between the US and Taiwan, thereby supporting innovation in both places. They also underpin the development of more formal transnational relations such as joint ventures and partnerships. The result is a complex web of formal and informal relations between individual investors and entrepreneurs, small- and medium-sized firms, as well as divisions of larger firms located on both sides of the Pacific. This transnational community is only partly constituted by the Taiwanese community in Silicon Valley and their 'home' connections, for example providing information on products and markets for firms in Taiwan.

There are two further critical elements. First, there have been increasing numbers of returnees to Taiwan who have become an integral part of the technology industries of Hsinchu. Many immigrants have returned to Taiwan to start businesses there while keeping close links with former colleagues and classmates in Silicon Valley. The Taiwanese government has actively encouraged return migration since the mid-1970s, a stance formalized through the establishment of the Hsinchu Science-based Industrial Park (HSIP) to provide land, tax incentives and infrastructure for returning investors. These connections have also been consolidated by the emergence of new bridging associations such as Monte Jade Science and Technology Association set up in 1989. Return migration grew appreciably in the 1980s (200 engineers and scientists per year on average) and accelerated in the 1990s (1,000 per year mid-decade). By 1997, Hsu and Saxenian's (2000) research reveals, the HSIP had attracted 2,850 returnees who accounted for the creation of 97 businesses in the Park, 40 per cent of the total number.

Second, there is a growing population that works in *both* regions. With families based on either side of the Pacific – most commonly in California due to lifestyle factors – these engineers travel between the two areas as often as once or twice a month. They include Taiwanese angel investors and venture capitalists as well as managers and engineers from firms with operations in both regions. These are some of the iconic ethnic Chinese 'astronauts' described by Ong (1999: 127) who also characterizes their domestic situation as 'families in America, fathers in midair', a clear allusion to the huge predominance of male migrants in these sectors. Together, these three overlapping and fluid groupings – the US-based engineers, the Taiwanese returnees and the 'astronauts' – constitute what Hsu and Saxenian (2000: 1999) describe as 'the bridge between Silicon Valley and Hsinchu'.

While the Taiwan-Silicon Valley transnational technical community is perhaps the most longstanding and developed in the Asia-Pacific, returnees from the US are beginning to establish other such IT-based communities in the region. Indians educated in America have played a key role in setting up Indian software facilities for leading US firms such as Oracle, Novell and Bay Networks, while others are setting up their own companies upon their return. Similar processes, albeit on a much smaller scale, are now discernible, for example, between Silicon Valley and Korea, Singapore and Vietnam. These transnational flows are not simply an addendum to the economic development of Silicon Valley and the other national IT industries concerned, they are an integral part of it. In an increasingly global economy, immigration, trade, investment and economic development are intertwined in complex and, in the case of Silicon Valley, mutually beneficial ways. Our key argument in this chapter, therefore, is that the embryonic transnational communities briefly described above both shape and are shaped by the pre-existing structures – economic, political, cultural and spatial – of the global economy in a mutually constitutive manner (see also Gertler and Amin, in this volume).

We conclude this section by making three points in this regard. First, the development of these transnational communities is not purely to do with the activation of ethnic networks. Taiwan–Silicon Valley connections, for example, cannot simply be explained in terms of Chinese business networks and *guanxi* relations. As Hsu and Saxenian (2000) argue, these links are based upon rapidly constructed and constantly monitored professional networks, supported as much by common educational and professional experiences as cultural ties. While information may flow freely through ethnic networks, that is not to say that firm capabilities will necessarily develop. In other words, market information and firm capabilities must not be conflated through 'over-socialized' interpretations that overlook managerial and technical expertise.

Second, we must not conceptualize this developing web of trans-Pacific connections as 'deterritorialized' or 'stateless'. As Ong (1999) cogently desc-ribes, while international managers and professionals are adept at processes of capital accumulation, they do not operate in 'free-flowing' environments, but in arenas that are controlled and shaped by nation-states and financial markets. Such state regimes and markets are constantly adjusting to benefit from the global nature of contemporary capitalism. State initiatives on both sides of the Pacific to promulgate IT-based transnational communities are a case in point. These then are very much place-based relations operating within, yet starting to change, the pre-existing matrix of state-business relations. Moreover, these trans-Pacific flows are contributing to the regionalization of economic activity and the emergence of incipient forms of regional identity – in this case in the Asia–Pacific region – that are notable features of the contemporary global economy (Dicken, 1998a).

Third, and relatedly, for all the talk of engendering 'mutually beneficial' transnational relations, we must not downplay the uneven power relations that characterize the contemporary global economy. While high-tech nodes on both side of the Pacific may benefit in economic development terms as these transnational communities develop, we must not lose sight of the fact that the industry is being driven by the seemingly insatiable innovative capacity of Silicon Valley and demand from the huge US marketplace. The recent downturn in America's technology industries (2000–2001) and the almost immediate effects felt in technology communities across Asia make this last point painfully clear.

CONCLUSION

Peter Dicken's research provides a powerful analytical framework for revealing the key structures and actors of the global economy (see also Yeung and Peck, in this volume). More specifically, he describes a highly spatially and organizationally variegated system of networks, characterized by increasing levels of international functional integration and predominantly shaped by the interactions between nation-states and transnational corporations at a variety of scales. Our aim in this chapter has been to explore the linkages and overlaps between this conceptualization of the global economy and those dealing with emergent forms of transnationalism. The points of contact between these approaches are numerous, encompassing migrant workers of varying skills and status, investment flows through socio-cultural networks, remittance transfers and innovative forms of transnational entrepreneurship. These networks clearly have important impacts on both the sending and receiving localities, helping to redefine the very socio-cultural constitution and economic base of those spaces.

Our argument is that these increasingly significant transnational networks and spaces are shaped by, and in turn reshape, the institutional, corporate and market structures of the global economy in a mutually constitutive manner. Put simply, globalization and transnationalism emphasize different facets of the same profound geographical restructuring of economic activity. For example, it is impossible to comprehend fully the restructuring of Vancouver and Toronto without understanding how social networks linked to Hong Kong have channelled significant property investment into those cities. Similarly, one cannot understand today's Silicon Valley without an appreciation of the activities of skilled Asian in-migrants and the emerging networks of trans-Pacific 'returnees' and 'astronauts' that tie the Valley to technology commun-ities across Asia.

We conclude by offering four comments as to how a much-needed dialogue between scholars interested in globalization and transnationalism might fruitfully proceed. First, as already mentioned, network approaches offer an epistemological and methodological common ground. Focusing on the constitution of socially and culturally embedded transnational networks and the forces that create, shape, resist and truncate those networks allows exploration of the crucial intersections between structure and agency. The aim here should be neither to identify globalization from above or globalization from below, but rather a sense of globalization 'in action', 'in performance'. In short, a focus on transnational practices and networks must 'embed the theory of practice within, not outside of or against, political-economic forces' (Ong, 1999: 5).

Second, it is clear that research needs to retain a focus on the elite networks of the transnational capitalist class that play such a disproportionate role in shaping the contours of the global economy – a recommendation Sklair (2001) has been making for some time. While scholars working under the transnationalism banner have contributed greatly to considerations of how emergent forms of transnational communities should be conceptualized and investigated, few have applied themselves to the elite networks that in certain ways define the 'rules of the game' within which other forms of transnationalism are generated.

Third, research that integrates notions of globalization and transnationalism will self-evidently have to keep the reworked nation-state in centre stage, thereby resonating with a key theme in Peter Dicken's work. The various forms of transnationalism we have discussed in this chapter have all been created, regulated and scripted to a large extent by the (sometimes competing) demands of nation-states. As such, this work can be used to counter interpretations that seemingly identify transnational corporations as *the* key power brokers in the global economy (see also Dicken, in this volume).

Finally, the complexity of these transnational economic systems poses fundamental challenges to researchers. In particular, work that brings globalization and transnationalism together will need to transcend a variety of disciplinary and geographic boundaries. Collaborative research is needed that not only spans multiple strategic (in geographical and institutional terms) sites within the global economy, but also brings together scholars working in economics and political science, for example, with those working in sociology and anthropology. Our feeling is that economic geographers, following the integrative spirit of Peter Dicken's work, can play a central role in this process.

NOTES

1   Recurring patterns within Chinese family firms, including the Chinese family firms that have engaged in property development in Vancouver and Toronto, include the proliferation of enterprises (including listed firms) owned and managed by family members (especially males); a large number of sub-contracting operations characteristic of personalistic networks linking firms backwards to sources of supply and forward to consumers; relative high speed and flexibility within these networks.

2   It is important to note that the Canadian government permits dual citizenship and citizens can acquire non-residency status when working overseas (thereby lowering Canadian tax obligations to zero).

Chapter 4

THE MARGINALIZATION OF EVERYWHERE?:
EMERGING GEOGRAPHIES OF EMERGING MARKETS

Roger Lee

INTRODUCTION

One of the apparently more bizarre manifestations of 'a world turned upside down' – to use Christopher Hill's (1975) expression – after the events in New York on 11 September 2001, was the headline which appeared in the *Financial Times* just over three weeks later: 'Hot money flooding into Pakistan' (Luce, 2001: 3). The assumption that the potentially destabilizing effects of US-sponsored military action against Afghanistan would cause capital flight from assets denominated in the Pakistan rupee – already one of the weakest currencies amongst the emerging markets of the developing world – was itself turned upside down with the marked inflow of funds into Pakistan. Such 'counter intuitive flows' – as one official at the Pakistan central bank put it – were explained in part by the repatriation of money-based assets, possibly accumulated through tax evasion and so vulnerable – for entirely separate reasons – to the US's global search for funds associated with terrorist activities. There were other explanations too, associated more conventionally with the inflow of official aid into Pakistan. Whatever the cause, the Pakistan rupee appreciated by over six per cent against the dollar in the three weeks plus period.

This sequence of events illustrates many of the themes to be explored in this chapter. I am concerned with the iterative relations between dynamically uneven geographical development – manifest especially in terms of geographies of risk and reward – and geographies of investment (especially of equity and bond finance) in emerging markets. On the one hand is the insistent – and often dramatic – potential for the re-scripting of economic geographies arising out of financial globalization. This involves the continuous evaluation and switching of capital through financial markets and the consequent exposure of economic geographies to these assessments and, potentially very large, flows of capital. On the other hand, the very processes of globalization reveal the geopolitical fragility of an apparently global financial architecture, the geo-social and geo-cultural relativity of decisions taken in and through the apparent (but in practice only rhetorical) single-mindedness of financial markets, and the continuing significance of distinctive national and other local identities and geographies (Dicken, 1998a; Gertler, in this volume) in shaping the geographies of economic relations in a none-the-less globalizing world.

What follows is organized into five sections. The chapter begins with an outline of the arguments to be explored before moving on to consider the nature of risk and risk aversion in driving financial markets and shaping the buoyancy of investment in emerging markets. Section three considers the construction of knowledge in financial markets and the (re)switching of mobile financial capital. The emergence of emerging markets as a globalizing investment category constructed via processes of othering and de-othering is discussed in section four. The conclusion reflects on the novelty or otherwise of emerging markets and on the notion of the marginalization of everywhere.

ITERATIVE GEOGRAPHIES OF FINANCIAL FLOWS: UNDERLYING ARGUMENTS AND THEMES

Flows of finance within circuits of capital are orchestrated through globalizing financial markets. These are becoming, if unevenly, increasingly significant spaces of hegemony and regulation within contemporary capitalism.[1] These markets gain their hegemonic status from the indifference of mobile financial capital to location or to sector. Their hegemony derives from the construction of ever more sophisticated, if fragile, financial knowledge produced and used in increasingly geographically sensitized ways over a range of time horizons, from the instant to the long term, to evaluate one set of activities or locations against another. This growth in financial hegemony has the effect of forcing the mobility of total capital not only through the devaluation of locationally fixed capital, but also through the construction of a widening range of financial instruments through which capital may be mobilized and flows of capital switched from one place/investment to another. Such increased capital mobility destabilizes fixed capital and so feeds back to extend further – potentially, at least – the liquidity, scale and power of financial markets. Thus the processes of the (re)switching of capital take place both in response to, and in the construction of, geographically uneven development. The consequence is the marginalization of everywhere. This is the central theme of this chapter.

Figure 4.1 illustrates the composition and growth of international financial flows to developing countries over the past three decades or so. Notwithstanding the now commonly made argument (see Swyngedouw, 2000 for a critique) that levels of global financial interaction were at least as high in the late nineteenth century as they are today, the growing global reach of financial markets and the range of transactions involved in international investment reflect the diversity of knowledge used in the (re)switching of capital. However, despite their rapid growth in the 1990s, such material flows pale into insignificance when compared to flows of capital across the foreign exchange (FX) markets, the value of which increased from a mere 10 times that of the combined value of flows in trade and foreign direct investment in the early 1980s, to over 70 times – at almost $400,000 billion per year ($1,095 billion per day) – by the late 1990s (see Figure 4.2).

Although Swyngedouw (2000: 64) considers such financial flows to represent the '"real" myth of globalization', the practically critical question is: what are their consequences? The argument in this chapter is that their effects are real, substantial and geographically formative. The mere possibility of such large-scale financial movements and switching is a profound 'regulator' of financial behaviour and economic norms. It evaluates geographical difference through globalizing criteria and 'normalizes' economic, monetary and fiscal policies within and between states. At the same time, the growth in the overall proportion of investment (both direct and indirect) and output accounted for by international flows and financial and commodity capital, increases space-time compression and so intensifies competition between units of surplus-earning capital and the tendency to their increased mobility. This in turn extends – potentially and unevenly at least – the influence of financial markets in their formative expectations of increasingly non-local norms of profitability.

Although such processes promote the shift of the 'natural' scale of economic regulation upwards to international states like the EU (Lee, 1990), they involve less a 'jumping' of scales (Smith, 1984) than a continuous destruction and construction of geographies of accumulation and the continuous (re)negotiation of the geographies of state regulation over multiple and interlocking scales.[2] This emergence of new geographies of accumulation has the effect of incorporating geographies of state regulation into new configurations of spatial relations. A good example of this would be the 'dollarization' of certain Latin American national economies and the painful struggle – so apparent in the Argentinean crisis of 2001–2002 – across a range of scales that accompanies the insistent construction of these new geographies.

The continuous (re)switching of financial and money capital also reflects the geographical imaginations and experience of those working within financial markets and the interactive construction and exchange of knowledge – often over very short time frames – in which they are constantly engaged. The spaces across which this (re)switching takes place are thereby vulnerable to the limited understandings, evaluations, whims and self-interests of the agents and advisers of capital and are closely informed by the ways in which wider political, social and cultural geographies – as well as the understandings and representations of them – frame the financial reading of information about the potential trans-formation of economic geographies. There is not therefore a singular movement towards, or away from, globally informed or locally structured economic geog-raphies. Rather, they are iteratively – and hence dynamically – interrelated and constructed out of the mutually formative relations between global responses to the sub-global, and sub-global responses to the global.

In one sense, however, the influence of globalization *is* rather more singular, uniform and global. The implication of the foregoing arguments is that the growing significance of financial markets for global regulation – and the attendant naturalization of capitalism generated by this growth – leads to the marginalization of everywhere. Rather than merely shifting the contested relat-ionships between the 'local' and the 'global' in shaping circuits of reproduction

Figure 4.1 Net capital flows to developing countries by type of flow, 1970–1988 (in $ billion)

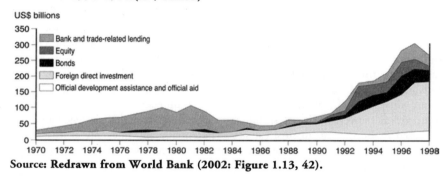

Source: **Redrawn from World Bank (2002: Figure 1.13, 42).**

Figure 4.2 Global capital, trade and foreign currency transactions (in $ billion)

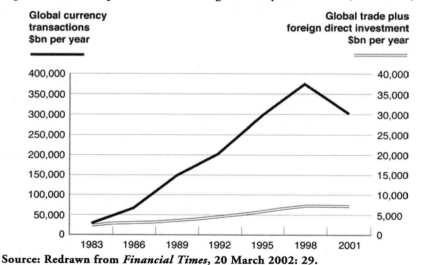

Source: **Redrawn from *Financial Times*, 20 March 2002: 29.**

or reconfiguring the contours of uneven development, the material processes of globalization (trade, investment, production and financial evaluation) and the intensification of the mobility of total capital increase the marginality of *all* such circuits and *all* the people and places engaged in, or disincorporated from, them. Zizek (1997: 44, cited in Swyngedouw, 2000: 74) refers to this process as 'the paradox of colonisation... the final moment ... of colonisation' is reached, he writes, when 'there are only colonies, no colonising countries'. Investment in emerging markets demonstrates not only this tendency towards universal marginalization, but also the associated and increasing levels of interdependence between particular localities and the constantly transformed scales of material reproduction within the globalizing economic geographies of contemporary capitalism.

FINANCIAL MARKETS AND GEOGRAPHIES OF ACCUMULATION AND RISK

Financial markets are markets in risk. They arbitrage between buyers and sellers who value risk (or, rather, the potential accumulative reward from certain levels of risk) in different ways.[3] The nature and quantitative significance of risk – as well as the social and cultural processes through which data on risk are transformed into usable and credible (but not necessarily rational) knowledge – are therefore defined in terms of the assessment of the prospects of reward. At its most simple, the higher the risk, the higher the reward expected. High risk assets command low prices – and/or a substantial risk premium – in compensation, so creating the possibility of high rewards.[4]

Thus markets in risk may enable the offloading of risky assets from risk-averse to risk-loving investors. Risk-averse investors prefer a smaller chance of higher returns in exchange for a reduced risk of lower returns. Risk-loving investors are willing to increase the chance of lower returns in exchange for the increased chance of higher returns. Markets in risk are thus highly dynamic. They are driven not only by changes in the magnitudes of risks and rewards over an infinite range of transactions and assets, but also by the often dramatically uneven spatial and temporal fluctuations in levels of risk aversion. Of course, changed levels of aversion may themselves reflect realistic responses to information and knowledge about risk and reward. Under such circumstances, it is difficult – if not impossible – to differentiate 'real' changes in risk/reward ratios from 'interpreted' differences founded on increased levels of risk aversion, and hence the substantial risk of 'herd' behaviour in financial markets and its even more substantial consequences.

The causes of these dynamics are complex. For example, the contagion between emerging markets as an asset class was far less apparent in the Argentine crisis of 2001/2 than it was in earlier crises (Mexico, 1994/5, south-east Asia, 1997, Russia, 1998). This was due in part to the relative unattractiveness of other investments in the low growth, low inflation environment in developed economies. But it was due also to the fact that investment in emerging bond markets in the early years of the new millennium was led increasingly by pension funds and insurance companies rather than by highly leveraged hedge funds that led investment in the later years of the old.

Emerging markets are especially vulnerable to such complex dynamics as they represent a cluster of asset classes with a relatively high risk, to or from which mobile portfolio capital may be attracted or repelled precisely because of the high rewards/risks that are priced into their purchase. Thus, emerging markets tend to be favoured by risk-loving investors, especially when the level of aversion to risk is low and are shunned – often dramatically quickly – when the level of aversion grows. Under such circumstances – aggravated since the events of 11 September 2001 – the tendency is for a flight to quality and the consequence is a reduction in portfolio flows to emerging markets. This makes such capital flows pro-cyclical and intensifies the effects of a slow-down in global growth on emerging markets. Since the south-east Asian crisis of 1997, for example, net investment into emerging markets has shown two divergent trends. Direct investment stopped growing but remained positive, whilst net portfolio investment dropped sharply into negative flows and, even by 2002, is forecast only to begin to move back towards positive net flows (see Figure 4.3 for data on all developing countries).

Figure 4.3   Net flows of investment into developing countries, 1991–2001, as a percentage of GDP

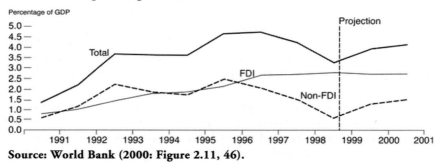

Source: World Bank (2000: Figure 2.11, 46).

Thus, advocates of portfolio investment in emerging markets in early 2002 pointed to the attractions of investment in 'unloved funds' (quoted in Burgess, 2002: vii). Reflecting the assessment of Kate Munday (head of emerging markets at Barings Asset Management) that emerging markets were cheap, with stock markets 'trading at a 40 per cent discount to developed markets' (which were themselves over-valued), Mark Dampier, of advisers Hargreaves Lansdown, argued that

> [T]here has been a seven-year bear market in emerging market equity [funds]. They are completely unloved, unwanted and unfashionable and that usually represents a buying opportunity.

As investments into emerging markets are, almost by definition, dominated by flows from more to less developed economies, such relationships represent a clear case of the development of under-development. The direction and timing of flows is influenced by calculations of risk/reward ratios by investors in the context of the diversification of these ratios in their investment portfolios. The developmental needs of the destination of the investment are therefore very much a secondary – if even a relevant – issue.

Such relations of development are well-exemplified by the evaluative power of portfolio investment managers and commentators over Africa (quoted in Gettleman, 1996):

> The presence of Africa funds means big chunks of money are out there which provide incentives for privatisation and put pressure on governments to improve their [macro-economic] policies. (Kader Allaoua, senior economist of the International Finance Corporation)

> African governments are beginning to realise that equity can be a great way to raise money. The region is finally starting to follow the rest of the world and investment attitudes are changing. (John Niepold, manager of the Washington-based Africa Emerging Market Fund)

> … all the signs of reform have been encouraging, and now's a good time to look for deals in African markets. (Mark Mobius, fund manager for Templeton)

The evaluative talk here is of making sure that potential destinations of portfolio investments in Africa move more quickly to create conditions attractive to such investment. And this power is more material than rhetorical. Bennell (1997) estimates that, in 1996/7, more than 800 state-owned enterprises in 17 sub-Saharan countries were being prepared for privatization and, on conservative estimates, over 1,500 privatizations would have been completed by the end of the decade.

DIVERSIFICATION, KNOWLEDGE CONSTRUCTION, EVALUATION AND CAPITAL SWITCHING:
THE FINANCIAL DISCOVERY OF GEOGRAPHY AND THE (RE)DISCOVERY, OR (RE)BIRTH, OF UNEVEN DEVELOPMENT

Diversification

Diversification (the holding of investments with unrelated returns) is the mantra of portfolio investment, although what it means in precise proportional terms (over what number of investment opportunities with what quantitative effect on the holdings of each) is rarely spelt out. It remains a principle rather than a template. Deviate from full diversification only with specific knowledge or with specific intent. The ability to diversify within an industry across countries as well as between industries within countries necessarily increases the range of possible strategies of diversification and serves to redefine the influence of the local/national distinction on global economic activity. Surprisingly, however, the frequent debates on diversification in financial media which, in labouring the point that diversification is scale dependent (although, of course, that term is not employed), reveal a stunted geographical imagination on the part of investors and their advisers. They miss the point that the difference between diversification within a national frame and diversification across a broader-than-national frame (usually referred to as 'global', although it is rarely that) is not merely one of geographical scale or of being able to invest in a sector in different national locations. Any increase in the spatial scale over which investments are made is, by definition, bound to influence the quantitative degree of their diversification. But an increase in the spatial range of investment also involves increased qualitative diversification resulting from the increase in the range, diversity – and scales – of the complex geographies encompassed by the investment.

Nevertheless, the simple distinction – national, multi-sectoral/global, uni-sectoral – frequently adopted in discussions of diversification, points to the continuing influence of the national on portfolio investment decision-making. Pension funds, for example, are typically more than 90 per cent domestically based. But, again, the distinction misses the point that all geographies are multi-scalar. Both 'national' and 'global' forms of diversification are founded on uneven geographies constructed at scales greater and smaller than the national in which, however, the national frame is highly influential. Nationally distinct regulatory regimes (and emerging markets might be considered to be a diverse categorical form of one such) are shaped, in part, by national responses to the effects of emergent international divisions of labour, international financial relations and the consequent constraints and opportunities on national economic performance and structure. But such international divisions of labour (which constrain the extent of intra-national sectoral diversity and hence the possibilities for the nationally based diversification of investment portfolios) are articulated through increasingly, but still far from dominantly, global exchanges of trade, finance and production. These global influences (including the intensification of capital mobility discussed above) on geographies of direct investment (domestic or foreign) and on corporate geographies, help to shape the possibilities of such diversification.

Knowledge construction, financial evaluation and capital switching

In dealing with risk, markets commodify uncertainty and unevenness. The agents of financial capital operating in such markets produce profit-making financial instruments designed to enable the offsetting of risks and uneven development inherent in any reproductive activity within circuits of capital, thereby also sustaining the possibilities of profitability. Thus, the formal construction of emerging markets as an investment category (see, for example, Sidaway and Pryke, 2000), and the financial knowledge industries that have grown up around them are themselves reflections of the codification of knowledge in the packaging of uneven development in order to enable its exploitation in globally diversified investment portfolios. The initiation or disruption of flows into or from emerging markets is dependent not least on access to knowledge about the investment possibilities presented by them, i.e. on the geographical range of knowledge that may be incorporated into a strategy of portfolio construction, the spatially and temporally variable assessments of risk/reward and the data/knowledge which facilitate such assessments.

However, notwithstanding the confident way in which commodified financial knowledge is represented as precise by its producers and sellers, the uncertainties surrounding these possibilities may be illustrated in assessments of the Russian market in the mid 1990s:

> Some [fund managers] view Russia as 'the mother of all emerging markets', stuffed with natural resources and 150m highly-educated people, which could develop into one of the world's great financial hot spots early in the next century. Others see it as an investment 'black hole' into which capital will simply disappear. (Thornhill, 1996: 24)

> The biggest task is simply trying to get buyers and sellers to meet. It is still very difficult to know what the price of a share is. (A 'frustrated western fund manager', quoted in Thornhill, 1994: 19)

The construction of such assessments takes place almost entirely through the multifarious interactions and information/knowledge exchanges within major global financial centres, which thereby pre-condition the identity of investment opportunities and the circumstances under which they may be taken to market. As a result, they evaluate emerging markets in terms which, as indicated above, reflect the objectives of capital accumulation rather than local development need.

Thus, if Figure 4.4 has any heuristic value, it can attain such only by bearing such processes of the social construction of knowledge and of assessments of risk/reward ratios in diversification strategies constantly in mind. What the figure attempts to show is that the evaluations of economic activity made in and through financial markets have the effect of increasing capital mobility. Mobilized capital then faces four possible alternatives: withdrawal – in full or in part – from circulation, recirculation/switching in place – but not necessarily in the same sector or range of securities – or switching into foreign direct or indirect (portfolio) investment. Such switching increases the intensity of competition and interdependence between localities and economic activities, which in turn further increases the mobility threat to existing spaces of capital.

Under these circumstances, 'dependent development' is an understatement: 'determined development' is a rather more appropriate description of the relationship between capital flows and their preconditions and local responses. Financial markets are far from being mere neutral clearing mechanisms for risk. Insofar as they work effectively, they serve to reduce risk – and so enhance accumulation – by creating and sustaining liquidity, and thereby facilitating exchanges between buyers and sellers. But, in articulating the formative interrelations between geographies of risk and geographies of accumulative reward, they become a powerful means whereby capital can most effectively seek out surplus value. In so doing, financial markets shape the geographies of flows of capital and so (re/dis)incorporate places and activities within/from circuits of capital.[5] As a result, in a world of neoliberal economic policy (see Tickell and Peck, in this volume), financial markets also become arbiters of economic normality and conformity founded, the intention is, unquestioningly and unproblematically on capitalist norms and values. And the point is that this evaluative arbitration is not merely of emerging markets, but of investment opportunities in general. Thus, even outside the false world of zero-sum games, a move into or out of emerging markets also means a move out of or into other investment markets. The construction of emerging markets as an investment category simply serves to intensify this marginalization of everywhere.

THE CATEGORICAL EMERGENCE OF EMERGING MARKETS

> We all lost friends in the towers, but you don't have to be the big bully in the street picking on the little guy still.

> This bullshit about the courts and legal process is total bollocks when we're talking about the Middle East because they're just so different from us. They broke all the rules by bringing that kind of terror campaign to totally innocent people. I think we should turn the place into a car park. (Respondents quoted in Branigan, 2001: 11)

Materiality and culture: the categorization of emerging markets

The emergence of 'emerging markets' as a significant and influential investment category (accounting for around five to eight per cent of private global investment flows) is far from the merely simulated construction of spaces of hyper-reality and, (*contra* Sidaway and Pryke, 2000) is certainly not 'strange'. Furthermore investment in emerging markets is hardly new (see, for example, Tudor, 2000). Nor is there anything new in the realization that discourse is the crucially formative context through which emerging markets – like all forms of economic interaction – are constructed (Ó Tuathail, 1997; Smith, 2002). Sure, the notion of emerging markets reflects poststructural rescriptings of global economic geographies beyond the banalities of 'first', 'second' and 'third' worlds (Sidaway and Pryke, 2000), or the imposition of geopolitical categories like shatter belts and gateways (see Smith, 2002 for a critique of their usage in Europe).

Figure 4.4 Circuits of capital and capital switching

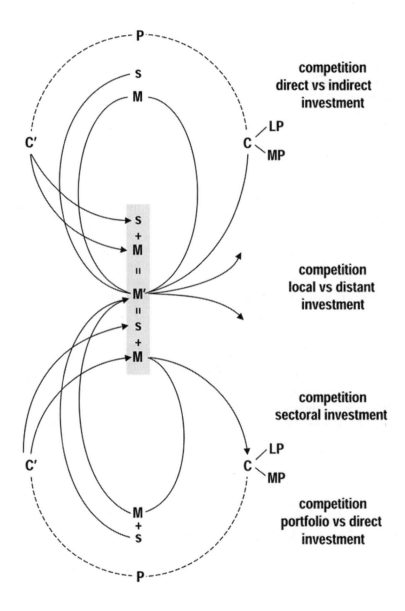

dominance of the norms of financial capital in shaping and categorizing global geographies.

Economic commentators, from Adam Smith (1893/1776) to Karl Polanyi (1944), sensitive to the social and political economy inherent to all economies, have always found it necessary to point to the mundane notion that economies are material processes discursively constructed out of social and cultural relations and understandings (Lee, 1989; 2002). Furthermore, from Buchanan (1935) to Buchanan (1970) and beyond, some economic geographers have set out to understand the objects of their concern from such a cultural/material perspective (see, for example, Lee and Wills, 1997: prologue and section 1). Thus, investment in emerging markets depends, amongst many other things, on the design, construction and availability of appropriate investment vehicles that involve the transmission of knowledge about investment opportunities in emerging markets to as large a number of potential participants as possible. But such vehicles must, above all else, themselves be profitable (i.e. be sufficiently effective in sustaining profitable investments in them) if they are to continue to be marketed. The continuing rediscovery of this kind of simultaneity of culture and economy in the social and discursive construction of the economic – illustrated in this chapter by the way in which the practical investment identity of emerging markets has been constructed – is therefore very much to be welcomed, even if the economic 'bathwater' must always be re-united with the socio-cultural and discursive 'baby'.

Emerging markets: othering and de-othering

The contemporary category of emerging markets and the criteria used to define them derive from the conscious efforts of the International Finance Corporation (IFC, a privately oriented affiliate of the World Bank) from the early 1980s to expand the world fit for capital investment. In 1981, the IFC launched the *Emerging Markets Data Base* (EMDB).[6] This was intended originally for in-house use, but in the face of increasing demand from financial capital for these indices and the underlying data on which they were based the IFC began offering its indices and data as a commercial product. The indices 'proved to be valuable performance benchmarks ... [and] have helped investors overcome the difficulties of comparing locally-produced indices with differing methodologies' (Standard and Poor's Emerging Markets Indices, 2001). S & P's current indices cover 54 emerging markets and 20 so-called 'frontier markets' that 'tend to be relatively small and illiquid even by emerging market standards' (Standard and Poor's, 2001).

The very terminology of 'emerging' and 'frontier' 'markets' reflects a powerful and formative process of economic othering (see also Smith, 2002). 'On the face of it', Tudor (2000: 7) writes:

> 'emerging markets' are easier to define by what they are *not* than by what
> they *are*. They are what is left when the world's major developed markets –

New York, London, Frankfurt, Tokyo, Paris, Milan, and so on are taken away.

But, of course, such a left-over world (see, for example Table 4.1 and Figure 4.5) would most definitely not specify a world fit for capital investment to anything like the appropriate level of geographical discrimination. Thus, as well as being 'others', emerging markets must also be non-others; they must conform to capitalist norms. As the IFC (1998: 2, also quoted in Tudor, 2000: 8) puts it in its *Emerging Markets Stockmarkets Factbook*:

> The term 'emerging market' can imply that a process of change is under way, with stock markets growing in size and sophistication, in contrast to markets that are small and stagnant. The term can also refer to any market in a developing economy, with the implication that all have the potential for development.

The sub-text here, of course, is that 'the potential for development' implies 'the potential for accumulation'. This accumulative potential is shaped locally by the provision of appropriate investment opportunities for financial capital (through, for example, the privatization of strategic assets) and the listing of securities in a functioning and appropriately liquid stock market. Cartographies of risk help to differentiate within and between emerging and frontier markets.[7] Beyond that, financial capital requires the translation of data about investment opportunities into knowledge about them, the interpretation of this knowledge by a favoured investment analyst, and an effective mechanism of switching capital. However, the 'potential for development' is always shaped not only by local, but also by non-local processes and social relations. These mutually formative intersections are built in – however imperfectly and unconsciously – to financial evaluations. This is a further demonstration of the formative mutual significance of multi-scalar economic and social process.

Table 4.1 Capitalization of world stock markets

Country	Percentage of the global total as of 30 January 2001
US	48.5
Developed Europe	31.3
Japan	9.7
Rest of Asia Pacific	5.3
Latin America	1.4
Rest of world	3.8

Source: *Financial Times*, 24 February 2001: 9.

Figure 4.5   The left-over financial world: capitalization of world stock markets as a percentage of global total ($23,492.4 billion, February 2002)

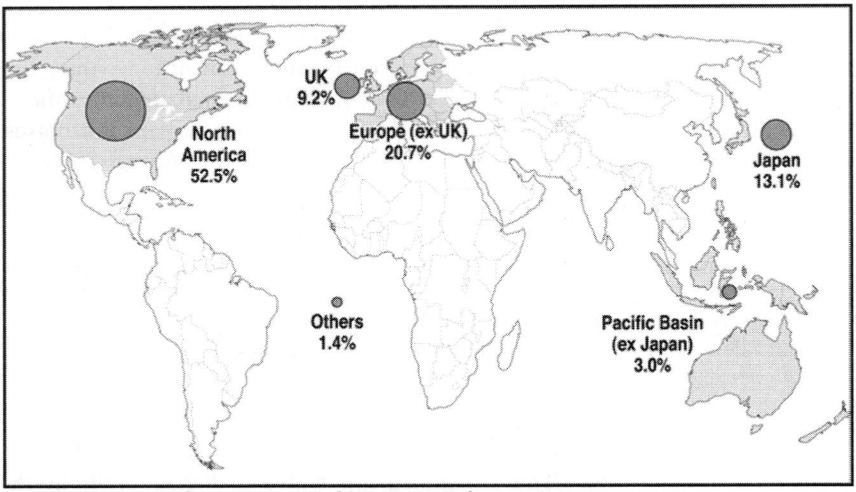

Source: Redrawn from *Financial Times*, 2 February 2000.

The significance of policy regimes in the emergence of emerging markets

The aptly-named 'Washington consensus' (apt, even though Thatcherism is airbrushed from history in its moniker) to replace the Bretton Woods system of economic regulation by that of (neoliberal and primarily financial) markets had provided the material and discursive context for the emergence of emerging markets. State controls over, and supervision of, short-term capital flows and fluctuating exchange rates were to be removed and multilateral financial and currency relations were to be encouraged positively to allow globalizing market forces to work as a continuously applied, but decentred, system of economic evaluation, judgement and discipline.

   In one sense, this development in policy – engineered by national policy makers, albeit through international institutions – merely followed material realities. The rapid expansion of the Euromarket in the period after 1973 created a large amount of mobile capital in search of investment opportunities. At the same time, an increased demand for capital from less developed economies – not least to cope with higher fuel charges – offset the rather poor returns available in the more developed regions. Syndicated loans came to dominate bank lending to overseas clients, often at negative levels of real interest. A turning point came in 1979 as monetarist policies based on interest rates to control inflation pushed up rates, so causing major problems for regions with loans denominated in dollars or other major currencies. At the same time, low commodity prices restricted the ability of debtors to repay the loans, whilst the debt crisis served to switch off capital flows to indebted regions.

Financial instruments and infrastructure in the emergence of emerging markets

The realization that debts would not be repaid simply because they could not be repaid provided the formative conditions for the emergence of the Brady Plan. This involved the restructuring of debt into so-called 'Brady Bonds' that discounted the face value of the debt or reduced interest rates and gave longer periods over which to repay the reduced debt. Brady Bonds were made attractive to investors by being backed by the US treasury bonds so that further default would not lead to a loss, merely a more limited gain. This financial revalidation of less developed regions made them attractive, in turn, to newly liberalized investors. Emerging markets were thereby encouraged to issue corporate or sovereign bonds – offering yields high enough to compensate for higher risk – and to seek ways of attracting newly mobile capital via programmes of privatization, the identification of key investment projects and the opening of stock markets. The effectiveness of these moves was enhanced by the relatively high rates of economic growth experienced by at least the more rapidly growing less developed economies. Such developments tended to generate a geographically desensitized contagion effect in what Galbraith (1994: 2) calls 'the financial mind' as a result of which most, if not all, less developed regions were presumed to have high or potentially high growth prospects.

However, like all markets, for investment in emerging markets to work requires constant infrastructural support and regulation. Financial instruments like American Depository Receipts (ADRs) were used increasingly to gain access to shares in emerging markets without risking exchange rate fluctuations or falling foul of restrictions on overseas investment in the destination countries. Specialist emerging market funds burgeoned, thereby enabling savings from a wide variety of sources to be switched with relative ease into investment in emerging markets. Especially important was the reliable and widely accepted upgrading of data into information and knowledge, provided largely through the *Emerging Markets Database,* about places and investment opportunities that are, almost by definition, largely unknown. This infrastructure of data/knowledge is particularly important in the case of emerging markets because they are constructed – orientalist-wise – out of what they are not, with lacunae defined almost exclusively via western and financial norms.

Macro-economic relations and the emergence of emerging markets

Lacunae are also all too apparent in market-based explanations for investment in emerging markets founded in two different time horizons and an inability to understand the essential geographies of economies. So-called 'structural explan-ations' are based on diminishing returns. They stress the uneven nature of global development and its effects on savings and investment rates. More developed countries, it is argued, have lower rates of growth and higher savings rates than less developed and the net effect is that capital is drawn out of more, and recirculated around less, developed economies in response to differential forces of supply and demand. In this view, investment flows to emerging markets are part of the equilibrating solution to uneven development proposed by neoclassical economics and articulated through efficient financial markets. The conclusion to be drawn is that such investment is a relatively long-term phenomenon.

Unfortunately, however, the data do not correspond to this simple picture. Divergence between groups of internally similar economies is what characterizes the contemporary geography of global uneven development. And this divergence goes back a long way – for at least 500 years (see, for example, Maddison, 2001). Since the late eighteenth century especially, rich countries have grown faster than poor countries leading to the well-known conjunction of sustained global growth and continuously increasing levels of inequality over the past two centuries (Lee, 2002). The geography of market-induced invest-ment flows is therefore rather more complex than that proposed by a simple two-region economic model.

An alternative macroeconomic explanation is based on the 'conjunctural' circumstances of low inflation and growth in more developed economies, equilibrated – in the short to medium term – by flows of capital to higher yielding emerging markets. Again, the presumption here is of a simple two-region model of the world economic geography driven merely by unproblematic forces of demand and supply and a single two-way channel of the flow of mobile capital. This hardly presages a space sufficient to accommodate the dramatic growth in the number of investment funds specializing in emerging markets (see Sidaway and Pryke, 2000 for some indication of the magnitudes involved), or at least not for their profitable coexistence.

Uneven development and the role of financial markets in the categorization of emerging markets

The drivers of portfolio investment in emerging markets are somewhat more complex than such mechanistic explanations, notwithstanding the value of them to those who wish to sidestep questions of power and exploitation in the promotion a neoliberal world based upon somehow socially unsullied market mechanisms (see Lee, 2002 for further elaboration). Investment may be enabled or disabled by a range of factors such as prevailing economic norms, knowledge of – and attitudes towards – risk and reward, and technologies for managing post-investment information flows and settlement systems. Local influences include secularization, the extension of market-oriented 'reforms' and market liquidity, and the provision of appropriate custody arrangements. All of these influences are multi-scalar in origin, definition and intersection; they cannot be reduced to the local or the global.

'Global shifts' of funds to and from emerging markets reflect not a simple set of core-periphery power relations, but the dynamic, if imperfect, under-standings and knowledges of geographies of risk/reward ratios and of assessments of the potential for change in the fortunes of an asset, all of which are highly susceptible to investment fashion. However, the effects of flows of investment – both material and discursive – are powerful and contribute towards the fallacious notion (see Lee, 2002) that with the increasingly widespread establishment of liberal capitalism, history has come to an end. Portfolio investment in emerging markets is yet another indication of capitalist incorporation. The uneven development of the Mexican and Argentine economies – and the associated social and political ramifications – during 2001–2002 illustrate these powerful effects.

Decoupling the other? A tale of two emerging Latin American economies

Mexico

Although emerging markets were widely touted as 'the investment fashion of the 1990s' (Coggan, 1993: vi), the devaluation of the Mexican peso in December 1994 underlined the fundamental problem of fashionability – its vulnerability to change. Despite US assistance to help avoid debt default, domestic economic policy to re-establish the credibility of the peso led to a severe recession in the mid 1990s. Notwithstanding the still continuing struggles of the Zapatistas to reclaim control over local circuits of reproduction in the Chiapas (see, for example, Russell, 1995; McIlwaine and Willis, 2002) – struggles which had unnerved the markets and so contributed to the national financial crisis – Mexican economic policy insisted on fiscal discipline and international integration, not least through the development of the North American Free Trade Agreement and the continued repression of alternative political and economic scripts. At the same time, an extension of the trade agreement with the EU was negotiated in 1999.

All of this has not only significantly changed Mexico's position within global economic geographies, but also has helped sustain internal political repression. Direct investment in Mexico is attracted not merely from North America, but from Europe and, in lesser degree, from Asia (Perez, 1998; Twomey, 2001): 'I think soon the country is going to be one big factory, a big manufacturing base for the NAFTA bloc' (Eduardo Cepeda, head of J P Morgan in Mexico, quoted in Tricks, 2000: 23). Such developments further disengage Mexico from emerging market status and, in March 2000, Moody's re-graded Mexico's foreign debt from Ba2 to Baa3, whilst Standard and Poor's re-graded the country's debt to one below investment grade. With such evaluative financial recategorizations, Mexico became increasingly decoupled from the effects of contagion emanating from the vicissitudes of emerging markets and, as a result, bond yield spreads were reduced.

But, by the same token, the country became far more closely linked to – and judged by – investment conditions characteristic of more developed economies. This linkage was also sustained by the election, in 2001, of Vicente Fox as president following decades of uninterrupted PRI rule. Thus, the economic disciplining of capitalism does not stop: the 'combination of fully-valued Mexican corporates (securities) does not make for a great risk reward [relationship]' (John Mullin, Latin American equity strategist at ABN Amro, New York, quoted in Mandel-Campbell, 2000: 30). Although Mexico is that much safer for capitalism, the threat of marginalization remains.

Argentina

This threat is all too apparent in the case of Argentina. Further slowdowns in inflows of portfolio investment during 2001 after a recession which, by the end of the year, had already lasted for 42 months, were not, unlike Mexico (or Brazil), offset by flows of direct investment. At the same time, the ability of Argentina to compete in international trade markets was further hampered by the link between the peso and the dollar. The continuing threat of devaluation and/or default on external debt therefore stimulated capital switching out of Argentine paper and equities, for example, and into Brazilian debt and equities. Internally, the consequences have been especially damaging. Severe restrictions were imposed on withdrawals from banks, an informal foreign exchange market developed (in which dollars were being offered at a premium of over 50 per cent above their pegged rate (one–to–one) against the peso), corporate defaults on debt repayment continued. Both formal and street-level violent political dissent against the austerity measures taken by the government led to a declaration of a state of emergency on 19 December. Along with the violence, looting and food shortages, this declaration – which brought the country's security forces under the direct control of the president – reflected the continuing threat to the legitimacy of the state – and most especially its government – posed by the crisis. The resignation, first of the economy minister, Domingo Cavallo, and then of president Fernando de la Rúa on 20 December, and the subsequent political turmoil revealed this loss of legitimacy in the most direct of ways.

The roots of this crisis go back to at least the mid-1980s and the IMF-induced restructuring of the economy involving the loss of many locally owned firms, state enterprises and jobs. Not surprisingly, a range of informal local economic geographies were constructed over the subsequent period[8] by those who had no choice but to continue to struggle to make a living, along with a multiplicity of local currencies – some issued in the form of bonds by provincial governments. One suggestion, made in response to the deepening of the crisis in 2001, was that such currencies should form the basis of a third currency – the Argentino – providing local liquidity (and hence, according to some critics, the likelihood of the return of inflationary tendencies which had dogged the economy before the pegging of the peso) alongside the peso and dollar. In the event, a wide range of quasi-currencies issued in the form of bonds by national and municipal governments, as well as privately issued and social currencies, emerged in the first few months of 2002. Estimates suggest that these quasi-currencies accounted for up to 75 per cent of the value of the 12 billion pesos in circulation (Catán, 2002: 12).

Responsibility for the crisis is closely disputed between local (in the form primarily of the middle classes who stand to lose most from a devaluation of the peso as their debts are denominated primarily in dollars), national (especially the established political elites), and international (the IMF) forces. Nevertheless, according to at least one influential commentator, the resolution of the crisis that underlay a dramatic expansion of the poor (in late 2001, one third of the population were unable to afford basic nutritional requirements and the rate of unemployment was around 20 per cent) must first involve the local:

> ... once the maximum politically feasible pain has been imposed on residents, the rest of it must logically be borne by foreign creditors. (Wolf, 2001: 20)

It is difficult to envisage a more direct example or hard-nosed description of the damaging formative intersection of local, national and global in the highly unsettling process of the continued making of histories and geographies in a world continually subjected to the marginalization of its reproductive conditions of existence.

CONCLUSION

Despite the hype, especially during the early to mid 1990s, the emergence of emerging markets is not such a novel story. It simply reflects one manifestation of the exploitation of uneven development by a capital made more mobile by the 'Washington consensus' around the hegemony of neoliberal markets. Furthermore, the possibility of the decoupling of emerging markets from such a status is not so novel either. It is, after all, commonplace in stories of regional uneven development that go on to chart the often-repeated engagement and disengagement of regions with, and from, extra-regional circuits of capital. The case of emerging markets points, however, to the mutually formative inter-section of multi-scalar relations. The crisis in emerging markets post-1997 had the effect of differentiating the representation of them by investment managers and advisers and led to, largely unheeded, calls for improvements to the 'global financial architecture'. On the other hand, the reaction of Malaysia to the crisis was to fence itself off, at least temporarily, from the most disruptive global circuits. In so doing, it earned the continuing opprobrium of neoliberal financial commentary. Similarly, the solution of the crisis in Argentina must depend on the emergence of a set of measures acceptable primarily to the middle classes that meets the 'regulatory' requirements of the IMF and the financial markets. Such a 'solution', however, must also address the politics of the material reproduction of labour and the distribution of value or be subject to more, possibly violent, resistance. But then it was ever thus – there are certain imperative links between the social and the material. It is, in other words, all too easy to fetishize emerging markets, seduced by the colourful language and exaggerated claims of the major movers and shakers working in, and profiting from, them.

However, three general conclusions suggest themselves from this brief survey. The first is the inadequacy of the global/local couplet in debates about globalization. Like all geographies, emerging markets are the constantly constructed and reconstructed products of a multiplicity of scales all working at once in a mutually formative and influential fashion. Second, the processes of uneven development are not reducible to simplistic core-periphery relations. The liberalization of capital flows in the period after 1979 pits all against all. 'Centres' and 'margins' are all marginalized, although the power of some localities to resist such marginalization is far greater than that of others. However, this is not such a novel observation either. A leader in the *Financial Times* published as long ago as 1995 (a time which might be regarded as the 'middle ages' of western interest in emerging markets) suggested that 'the day when markets in developed countries reflect similar risks to those in developing ones may not be far off'. This prediction was borne out – at least in the market for bonds – by 2002. According to Jerome Booth (quoted in Skypala, 2002: 8), head of research at emerging market debt specialists Ashmore Investment Management, 'The default risk is 10 times higher in the high yield corporate bond market than in emerging market debt.'

Finally, it is difficult to better the paraphrase of Marx's (1996: 90) comment on the mutual involvement of local and global scales of economic relations. This paraphrase appeared in Leys' (1994: 47) review of two books on what he called 'the African tragedy', published just as emerging markets were about to break more fully into the consciousness of the 'financial mind' in late 1994. Here, therefore, it is simply re-paraphrased:

> If, however, the German [or British or American] reader pharisaically shrugs his shoulders at the conditions of the English industrial and agricultural workers [or emerging market populations], or optimistically comforts himself with the thought that in Germany [or Britain or the US] things are not nearly so bad, I must plainly tell him: De te fabula narratur![9]

And this somewhat before the events of 11 September 2001.

ACKNOWLEDGEMENT

As ever, I am indebted to all those – including the editors of this volume – who constantly help me to try to understand what it is that I try to say. Peter Dicken remains the most remarkable and remarkably supportive teacher of a wayward student. Thank you, too, to Ed Oliver who produced the diagrams and my endless alterations of them with all the efficiency, expertise and tolerance that will forever amaze me.

NOTES

1   See, for example, Story (1999: 62–3) for an account of the way in which 'world financial markets [have], in effect, become judge and jury of the world economy'. For an historical geography of regulation through the imposition of western financial norms via the imposition of Structural Adjustment Policies, see Mohan et al. (2000) and Tickell and Peck, in this volume. The role of the World Bank in establishing the neoliberal framework for capitalist expansion is outlined by Cammack (2002) and a more general critique of the significance of financial markets is offered by Swyngedouw (2000).

2   I have to confess as to having some difficulties with the currently fashionable stress on scale that seems to me to be subsumed by the notion of (the always dynamic and complex) geographies through which economics and politics are played out. However, the insistence on the politics of scale in the literature (Smith, 1984, Brenner, 1988a; 1998b; Swyngedouw 2000) is an extremely important one in terms of recognizing the need for a scale-awareness in political struggle and resistance (see also Brenner, in this volume).

3   Accumulation is the definitive purpose of capitalist circuits of reproduction. So, to reduce accumulation merely to a reward for risk is therefore to indulge in the crudest form of commodity fetishism. It is to ignore the power relations involved in the extraction of surplus value and the range of institutional structures – both in markets and the state system – designed to facilitate accumulation.

4   This is the basis of the Emerging Markets Bond Index – a widely used measure of the relative attractiveness of investment in emerging markets. The index, produced by J P Morgan, calculates the spread of returns in emerging markets compared (normally) to interest rates on bonds in the US. The wider the spread, the greater the risk aversion affecting investment in emerging markets, and the greater the volatility of the Index, the greater the responsiveness of finance capital to perceived or actual fluctuations in risk.

5   It is precisely this perpetually reshaped but formative interplay of complex arrays of geographies that belies exaggerated claims about the death of geography in contemporary economic affairs (Lee, 2002).

6   The EMDB was acquired by Standard and Poor's in January 2000 – a transfer of ownership that reflects both the profitability of financial knowledge and its practical usefulness.

7   Merchant International Group is an example of a company that produces such knowledge in the form of what it calls Grey Area Dynamics™. It describes GAD™ as a 'killer application' in quantifying the geographies of risk. Pakistan emerges as the country with the highest 'overseas investment risk' according to the GAD™ index (see Morris, 2001: 14). The terminology and unreflexive positioning – to say nothing of the commodification – here are typical of the ways in which the geographies of emerging markets are represented.

8   The so-called Red Global de Treuque or Global Barter Network is one of the most highly developed and well-known of these local economic geographies. It was

established in Buenos Aires province but has since spread to provinces throughout Argentina. The network functions via a series of weekly meetings of each *nodo* or cell. It uses the internet (http://www.truqueclub.com; www.visitweb.com/treuque) as a means of articulating its economic geographies by disseminating information on the goods and services that may be exchanged (Pearson, 2001). On the construction of informal local economic geographies in a very different context see, for example, Lee (1996; 1999).

9   'The tale is told of you' (Horace *Satires*, book 1, satire 1).

Chapter 5

THE GLOBALIZATION OF ENVIRONMENTAL MANAGEMENT:
INTERNATIONAL INVESTMENT IN THE WATER, WASTEWATER
AND SOLID WASTE INDUSTRIES

Erica Schoenberger

INTRODUCTION

If we take the title of Peter Dicken's (1998a; 2003) book *Global Shift* seriously, it tells us two things. One is that things are constantly moving around the globe. The second is that the nature of this process of moving around is itself subject to change. If we take it as given that in a global capitalist economy, capital circulates internationally in a variety of forms (as commodities in trade, as financial capital in portfolio investments and loans, and as productive capital in foreign direct investments) and through a variety of sectors (raw materials, manufacturing, services, etc.), it is nevertheless the case that the kind of activity engaged in and the form the investment takes evolve over time as conditions change. Why these changes occur and how they play out socially, economically and geographically matter, and so it is part of our task to keep track of them – as Peter has shown us.

Over the last two decades, a new field of international direct investment has opened up. Most broadly described, it encompasses international investment in various aspects of urban and industrial infrastructure and the management of environmental inputs and wastes. The World Bank, which is not known for the idle use of the term, describes the advent of this new arena for global private investment as 'a revolution' (Sader, 2000: 1).

The proximate reason for this revolution is the wave of privatizations that were inaugurated by Mrs Thatcher in the UK and became part of a globalized orthodoxy of liberalization during the 1980s, enforced by the strictures of multilateral financial institutions. The state withdrew from a whole range of activities that had, until then, been considered 'naturally' part of its ambit either as direct producer or as close regulator of private monopolies, and turned them over to private enterprise and international competition.

This chapter considers the peculiarities of these sectors and the experience of international investment so far, with particular reference to investments in activities concerned with environmental management such as the provision of water, wastewater treatment and solid waste management. It will show that these sectors closely mimic early rounds of international direct investment in manufacturing. Specifically, the investments are geographically concentrated in the advanced industrial areas despite huge investment needs in developing countries. They are also largely undertaken through acquisition rather than greenfield investment. Industry consolidation and both vertical and horizontal integration are the guiding strategic mantras and this is resulting in a highly concentrated form of international investment.

The broad analytical framework I shall rely on aims at a synthesis of Harvey's (1982) 'spatial fix' with Peck's (2000) 'institutional fix'. The spatial fix concerns the ability of the capitalist system to absorb surplus capital through geographic restructuring. We usually think, in this context, of the relocation of investment inter-regionally and internationally and this is part of the story here. But the spatial fix in this case also includes the articulation of private capital with the physical environment and the production of urban and industrial infrastructure in a new way. This new way has been opened up by the rearticulation of the state with the private economy – Peck's institutional fix – as it has opened up some of its traditional orbit of activity to private capital. This combination will allow us, I think, to understand the underpinnings of this investment boom and its geographical and temporal characteristics.

WHAT IS AT ISSUE

The privatization of public infrastructure and environmental services has affected a broad range of activities from telecommunications to energy supply, road building and other transportation infrastructure, pollution abatement and control, solid and hazardous waste management, water supply and wastewater treatment. Private companies have long participated in these activities, of course, as suppliers of equipment and services and building contractors. The key difference is that they now also increasingly own the physical assets and manage them with a view to obtaining an acceptable rate of profit on the whole operation.

It is remarkably difficult to ascertain the size of the industries involved or their degree of internationalization either in the form of exports of products and services or foreign direct investment. I want to focus here on those businesses related strictly to environmental management, including water supply, wastewater treatment and waste management. The statistical data that are available concerning the size of the relevant global markets and industrial organization within these markets are fuzzy.

This is partly due to unreliable data gathering and reporting at the national level, partly due to the fact that

1   private investments in environmental management (e.g. investments made by manufacturers to control effluents) are not necessarily reported separately and so their magnitude is quite uncertain;

2.   many market participants are highly diversified, so that it is hard to disentangle their environmental businesses from other activities;

3   every data provider has a somewhat different definition of the industries involved (e.g. some include investments in 'clean' and alternative energy generation and the like).

The OECD estimates the global environmental services market for 2000 at $300 billion. This estimate rates all of Latin America, including Mexico, and the developing areas of East and Southeast Asia (including the newly industrialized economies) at zero without explanation. While it is plausible to suppose considerable under-investment relative to need, it is not the case that there is *no* investment in, e.g. water supplies in these countries, so this global estimate must be considered very conservative indeed. The US Department of Commerce, for example, anticipates that the Latin American market for the water and wastewater sectors alone will be $27.4 billion over the next five years. A private group, Environmental Business International, estimates that the water and wastewater sectors in emerging markets in Asia amounted to roughly $9 billion in 1996. In the same region, the World Bank anticipates water and sanitation investments of roughly $150 billion over the period 1995–2004. Only a portion of this will be entirely privately owned, of course, but a number of Asian developing countries, including Malaysia, the Philippines, Indonesia, China and Thailand are outsourcing water supply and wastewater treatment at a growing rate (*US Water News*, 1997). Nevertheless, as we will see, the main markets for investment are developed countries, especially the US and Western Europe.

With a somewhat more expansive definition of the whole environmental management and services industry than the OECD, the US government estimates the global environmental market at $513 billion in 2000. A European Community study estimates the global market for environmental goods and services in the range of 330 to 410 billion euros in 1999 (OECD, 1996: Table 1; US Dept of Commerce, 2000: 20–8; European Commission, 1999: 5).

Because the statistical data provide such an indistinct view of these activities, it seems worthwhile to look more closely at the operations of a particular firm to get a better sense of what the data might mean. The international environmental business has had some notable ups and downs in recent years, from which we can suppose that the path to profitability has been more difficult than imagined at the outset. Texas-based Enron, a large power supplier, expensively entered into the international water and wastewater business through the acquisition of Azurix and then abruptly exited when it turned out the synergies between power supply and water supply were less than anticipated. Meanwhile, it has, of course, been famously involved in a long-running lawsuit over a power supply deal with the Indian state of Maharashtra and has recently and spectacularly filed for bankruptcy.

Another American firm, Waste Management, Inc. was until recently one of the largest international players in the solid and hazardous waste business, with operations in the US, Canada, Mexico, Europe, the Pacific Rim and South America. Revenues from the international business amounted to $1.6 billion in 1999 out of total group revenues of $13 billion. The company then entered into a period of financial and legal crisis (including a class-action investor lawsuit and a formal SEC investigation) and has since sold off most of its international assets (Waste Management, 2001a; 2001b).

Here I want to focus on the French firm Vivendi Environnement. It recommends itself to study for its size and its recent, intensive international growth trajectory. While not in any sense statistically representative, it may be fairly considered emblematic of the changes taking place in the environmental services industry.

VIVENDI: AN OPERATIONAL PROFILE

In the water and waste businesses, at present, the main centres of power and international expansion are located in Western Europe. The top three firms in the sector are all European: Suez Lyonnaise des Eaux (France), Vivendi Environnement (France) and, since its recent acquisition of Thames Water (UK), the German power utility RWE AG (*US Water News*, 2000a). The smallest of the three, RWE/Thames, has a group market value of 20 billion euros and a total of 34 million water customers in Europe, the Americas, the Middle East and Asia.[1] Suez was, until recently, the clear global leader in the provision of water services. Its water division alone has annual revenues of roughly $8 billion and operates in 120 countries, serving 110 million consumers and some 60,000 industrial clients. The Group as a whole has annual revenues of $33 billion (*US Water News*, 2000b; Suez Lyonnaise des Eaux, 2001).

Vivendi Environnement is smaller overall than Suez, but in the last year or two has emerged as the largest provider of water and wastewater treatment infrastructure, products and services in the world. It is also involved in solid waste management, urban transport and power supply. Through its Spanish affiliate FCC (28 per cent owned by the company), Vivendi is involved in construction, waste management, water supply and cement production in Spain and Latin America.

The core group was originally an old-line, technically capable water company called Compagnie Generale des Eaux, founded some 150 years ago. Generale became part of a huge conglomerate, renamed Vivendi Universal, which has become a major presence in the media and entertainment businesses. Vivendi Environnement was partially spun out of the conglomerate at the end of 1999 and has been separately quoted on the Paris Bourse since 2000. Vivendi Universal presently owns 63 per cent of its environmental services affiliate (Vivendi Environnement, 2001a: 4811; *Financial Times*, 2001).[2]

The environmental group as a whole has been growing at a rapid clip, increasing its revenues by some 73 per cent over the period 1998–2000. This growth, as might be imagined, was achieved largely through acquisitions, especially in water/wastewater treatment, solid waste disposal and via its Spanish/Latin American subsidiary.

Table 5.1 shows the sectoral distribution of Vivendi activities. In 2000, Vivendi's water business had revenues of 12.9 billion euros, nearly doubling from 6.8 billion in 1998. Water supply, wastewater treatment and other water-related products and services such as filtration equipment dominate the business, accounting for just under half of total revenues in 2000, up slightly from a 44 per cent share in 1998. The 145 per cent growth in FCC, the Spanish/Latin American subsidiary, was also probably heavily committed to water, although detail is lacking. The other major growth sector, gaining by 82 per cent over the two-year period, was waste management.

Table 5.2 describes the geographic distribution of investment, again showing the change from 1998 to 2000. Unsurprisingly, the largest share of the business is in France (where Vivendi controls roughly 50 per cent of the water supply market), and the growth rate over the two years in question has been a relatively modest 20 per cent – although 10 per cent annual growth in a mature market is not to be sneezed at. The company is also well established in the UK where growth was a bit slower at 16 per cent. These two countries accounted for a bit over half of total revenues in 2000, but this was down from 76 per cent only two years previously. So it is quite plain that the company has focused on geographic expansion and diversification in growing the business.

Table 5.1   Vivendi Environnement's revenues by sector

Revenues (millions of Euros)	2000	1998
Water	12,856.2	6,758.5
Waste	5,260.0	2,897.0
Energy	3,220.8	2,739.0
Mass Transit	3,063.0	2,041.2
FCC	2,080.9	848.0
Total Revenues	26,480.9	15,283.7

Source: **Vivendi Environnement (2001a: 4825).**

Table 5.2   Vivendi Environnement's revenues by region

Revenues (millions of Euros)	2000	1998
France	11,081.0	9,319.9
UK	2,713.2	2,344.0
Rest of Europe	5,485.8	2,476.9
US	4,871.0	519.0
Rest of World	2,329.8	623.9
Total Revenues	26,480.9	15,283.7

Source: **Vivendi Environnement (2001a: 4825).**

In Europe outside of France and the UK, the company's operations grew by 120 per cent over two years. In the Rest of World outside the North Atlantic developed economies, the position grew by 270 per cent. But the most impressive growth of all was realized in the US market, where revenues grew by more than a factor of eight. Granted, this was from a relatively small starting position, but by the end of the period, Vivendi had become a serious presence in the US market with annual revenues of 4.9 billion euros. This is entirely apportioned between water (3.7 billion) and waste management (1.2 billion). By the end of the period, the US accounted for 18 per cent of the company's total revenues, up from three per cent.

In sum, in a two-year period, the company has restructured geographically to an astonishing degree. It has done this entirely by growing in new areas rather than disinvesting in old markets; indeed, it has continued to grow in its traditional markets. Good access to capital has allowed it to go on what amounts to a buying spree in its target growth markets, snapping up, as we will see, very well positioned firms in desirable market segments. Indeed, many European environmental service providers, benefiting from the sizeable cash flow of regional or local monopolies, find themselves in an excellent position for acquisitions (*Financial Times*, 2002a).[3] Although the geographic restructuring has been quite diversified and includes important forays into developing area markets, the overwhelming focus has been on expansion into already highly developed industrial and urban markets in Europe and, most particularly, the US. Note that these markets are characterized by a high degree of indigenous competition and relative maturity so that the markets as a whole are not expected to grow rapidly, although Vivendi's position in them has. And, of course, one of the best ways of dealing with local competition is to acquire it, which has certainly been one of Vivendi's strategies along with technology acquisition – or, at any rate, the acquisition of technologically innovative firms.

The overall, long-term strategy of the company is to offer integrated environmental management services combining, e.g. water, wastewater treatment, solid and hazardous waste disposal, etc. in a one-stop shopping style (Vivendi, 2001b). At a guess, one would suppose that the industrial market is more ready for this kind of integrated management package than urban or public sector markets generally, if for no other reason than their more consolidated decision-making. In any case, the company's own public relations on this point feature references to Alstom and Renault as model clients. Moreover, with Vivendi's enhanced global reach, it can partner other international firms wherever they are likely to set up shop. This hints at the likelihood that international environmental management investments will continue to shadow the geographical distribution of manufacturing and energy production.

The company's strategy statement underscores the desirability of focusing on 'high value added environmental services', which also suggests a continued strong focus on developed country markets. It may be stressed here that although environmental services such as water and wastewater appear to be rather ordinary and humdrum background features of modern life, the business is in many ways quite heavily involved in technology development and implementation. In this context, 'high value added services' are not simply an artefact of, say, market power, but also reflect a considerable level of technological sophistication and the investment and technology rents that go along with it.

Vivendi Water

Vivendi's water businesses worldwide, known as Vivendi Water, include a number of activities. First and largest is municipal water supply and wastewater treatment. In the US, its customers include the City of Chicago. In France, of course, it is ubiquitous, and it is also strongly entrenched in the UK through its wholly owned subsidiary, General Utilities PLC and that company's sub-sidiaries. Elsewhere in Europe, it has wholly- or majority-owned affiliates in Germany (Berlin and Leipzig), Hungary, and the Czech Republic and various other holdings on the continent. Like other international water companies, Vivendi is also a contractor to large industrial concerns such as GM, British Petroleum and Conoco for water treatment facilities and services. The company also manufactures and sells water treatment and filtration equipment and chemicals to the consumer, municipal and industrial markets.

In the Rest of World outside Europe and North America and apart from FCC's operations in Latin America, Vivendi's water business amounted to 1.2 billion euros in 2000, up 170 per cent from 445 million in 1998 (Vivendi, 2001a: 4825). This growth does not appear to have been accomplished by the acquisition or implantation of partially or wholly-owned subsidiaries (at least, none are listed in the company's annual report). So we must suppose that these are contract revenues recorded by the parent or its European or North American subsidiaries (Vivendi, 2001c).

Vivendi's rapid growth in the US water business occurred principally via the acquisition of a California-based company, US Filter, with North American sales of $4.5 billion in 1999. US Filter develops and manufactures quite sophisticated water and wastewater treatment products such as the eponymous filters, along with more prosaic products such as valves, meters and pumps. It also acts as a systems engineer for the design, construction and management of municipal and industrial water and wastewater treatment without itself necessarily owning the facilities. Its 'public-private municipal partnerships' include contracts with some 150 towns and cities in the US and Canada, including Petaluma, California and Atlanta-Fulton, Georgia. In the industrial outsourcing market, it is a contract provider at some 300 plants run by companies such as Chevron, VLSI, NEC, GM and Ford. The company describes itself as the largest competitor in the US in each of its four business groups. It spends about 1 per cent of its annual revenues on research and development, which is not terribly high at first glance. Although, if it is, as one might anticipate, concentrated in its more high-tech business lines, then its R&D effort could be a substantial proportion of those sales (US Filter, 2001).

As with its parent, US Filter has staked its strategy on integration of products and services in its various businesses. And, as with Vivendi, it has achieved its ability to operate in this fashion through acquisition. In the company's own words, 'US Filter began consolidating the industry' to overcome what it saw as traditional problems of fragmentation and hyper-specialization in the business which prevented the attainment of scale economies and meant that no one had the ability to offer 'total water manage-ment solutions'. Between 1990 and 1999, US Filter acquired a whole string of companies in different segments of the water business, becoming 'North America's largest and most influential water and wastewater treatment company', before being acquired itself by Vivendi in 1999. This last consoli-dation has afforded US Filter a stronger and broader international platform for growing its business (US Filter, 2001). At the same time, it has allowed Vivendi to become a major presence in the US market virtually overnight, while giving it access to state-of-the-art technology in certain segments of the business.

Vivendi's second major investment in the US water industry has been the acquisition of a 17 per cent share of Philadelphia Suburban Corporation (PSC), one of the largest investor-owned water utilities in the US, with revenues of $275.5 million and a market capitalization of over $1 billion in 2000. The company serves some two million residents in Pennsylvania, Ohio, Illinois, New Jersey, Maine and North Carolina. Vivendi is PSC's largest shareholder. PSC is a company very much in the contemporary image of Vivendi, pursuing growth through acquisition of municipal and private water systems in the US (PSC, 2001; UBS, 2001). As with US Filter, the company's watchword is 'consolidation' of an excessively fragmented industry.

Waste Management

The umbrella group within Vivendi for waste management and transport is CGEA, Compagnie Generale d'Entreprises Automobiles. Its principal subsi-diaries are Connex for road and rail transport services, and Onyx for municipal and industrial waste collection and treatment. Here I want to focus strictly on the waste management side of the business, but as revenue figures and other quantitative data are available only for CGEA as a whole, this requires some crossing back and forth between the different activities.

As with Vivendi as a whole, CGEA has been growing rapidly and gaining especially in international markets. Group revenues tripled from 1.8 billion euros in 1995 to 6 billion Euros in 1999. The international share of the total was 18 per cent in 1995 and 52 per cent by 1998 (figures for 1999 were unavailable). Total employment in 1995 was roughly 43,000, of which one third were international. In 1999, total employment was up to 131,000, of which almost 70 per cent were international (CGEA, 2001). Once again, we see a pattern of tremendous growth and geographical expansion.

Waste management appears to have been the main focus of this expansion. In 1998, Vivendi's waste management operations brought in 2.9 billion euros compared with 2 billion for transport. By 2000, waste management revenues were 5.3 billion euros, up 82 per cent. At least part of this increase included the acquisition of some of the international assets spun off from troubled Waste Management, Inc. during the same period (see above). Transport revenues had increased to 3 billion euros, up just 50 per cent. Still, by any standards, the two-year growth rate is remarkable and reflects, again, an aggressive strategy of acquisitions in target areas.

The chief growth target area for the waste management arm was, again, the US. Here revenues grew by over 1,000 per cent over the two-year period, reaching 1.2 billion euros in 2000. Revenues in Europe outside of France and the UK grew by 79 per cent, while in the Rest of World they increased off a small base by some 340 per cent to 655 million euros. By the end of this period, then, waste management revenues derived principally from France (45 per cent of the total, down from 64 per cent) and the US (23 per cent of the total, up from 3 per cent) (Vivendi, 2001a: 4825).

Analysis

The first thing we can underline is a tremendous and vary rapid push outwards from the country's home base in France and, secondarily, the UK. This has meant, in the first instance, expansion into the Rest of Europe, especially other European Community countries, but also some Eastern European states that are in line for accession. Nevertheless, the share of Europe as a whole (including France and the UK) declined from 92 per cent to 73 per cent of Vivendi's total revenues from 1998 to 2000.

Second, we can see that despite the attention focused on privatization and growth in emerging markets (see Lee, in this volume), especially in Asia, developing countries as a whole have not been a major investment target for Vivendi. If we take Rest of World as a proxy for developing countries, their share of total revenues has increased from 4 per cent in 1998 to 8.7 per cent in 2000 (see Table 5.2). In fact, Rest of World also includes developed Pacific Rim countries such as Australia, so these figures probably somewhat overstate the case. But here is what they suggest. The big international companies are paying attention to the developing world, but they are doing so cautiously. Vivendi has approximately doubled its apparent exposure in these regions, but this still accounts for a small share of its total efforts. At a guess, this small share is concentrated in a reasonably small number of countries, so the local impact of Vivendi (and other companies in the business) in certain developing countries may be quite high. But from the company's point of view, the developing world has not yet become a compelling market.

Why not? One can imagine a number of factors. First of all, the markets aren't wealthy or deep enough, especially following the 1997 financial crisis in Asia, although the *need* for environmental services is great. Unlike the providers of luxury goods, cars or even pharmaceuticals, environmental management companies may not be able to live off just the wealthy two per cent. They need a broader customer base to support the investment and yield local network and scale economies. In addition, manufacturing is not dense enough on the ground or facing sufficiently effective environmental regulations to support a large industrial market for firms like Vivendi. Operating costs could also be substan-tially higher despite incredibly low wages for local workers. These would include the cost of supporting expatriate managers and engineers, as well as generally higher costs for the acquisition and transport of material inputs. Finally, there aren't large, relatively straightforward acquisitions to be made in these markets. At a guess, the revenues from developing countries are derived mostly from contracts, which would count as exports, rather than from direct investment involving ownership of the underlying physical assets.

Third, the overwhelming focus of the firm's global repositioning is the US. This is both odd and reasonable. It is odd because the US must count as one of the more mature markets around, particularly for basic water supply, wastewater treatment and waste disposal, suggesting that market growth will be gradual. There are plenty of local competitors with well established presences in the market and long-term relationships with customers. On the other hand, the market, if slowly growing, is huge and well developed: small growth in a huge number is perhaps even more attractive than rapid growth off a small base. There are lots of excellent acquisition targets. This gives one instant entrée into a developed market, instant scale economies and a good outlet for the efficient investment of surplus capital on a large scale. Further, because of historical differences in regulatory regimes across markets, especially between Vivendi's home market and the US, an acquisition strategy (versus greenfield investment) is even more strongly indicated as a way of buying experience and appropriate technology.

The whole question of technology ownership and acquisition is a complex and interesting one in the field of environmental management. At least until recently, as indicated, the European approach to regulating wastewater treatment was quite different from that adopted in the US since the passage of the Clean Water Act in 1972. In brief, water supply and wastewater management in most European countries were organized around entire river basins and targeted ambient water quality standards. This system was agnostic on how these standards would be achieved; users could select the technologies they found most appropriate at each discharge point, provided the ambient water quality standard was met. The US, by contrast, mandated *discharge* standards and referenced these to a particular water treatment technology. In short, the discharge standard and the technology were both uniform – covering all situations nationwide no matter what the local circumstances were – and tightly coupled. The traditional European approach is widely viewed as providing substantially more scope for experimentation and innovation in wastewater treatment technologies, although it may suffer from other problems. The EU, however, has lately been moving towards the US model of technology-based regulation where it describes the technology that should be used in order to achieve the desired outcome (Marino and Boland, 1999). On the other hand, there is no specific reason to suppose that the EU reference technologies will be identical to those in the US.

The convergence of regulatory regimes between the two largest global markets for environmental management will surely make it more attractive and more necessary for firms to expand and deepen their presence on both sides of the Atlantic. This again augurs a wave of acquisitions, both to achieve this presence as fast as possible, and to cover the necessary bases in terms of technologies in use. Certain kinds of economies of scale and scope will be more available and more salient, sharpening the kind of 'consolidate or die' mentality that already seems to characterize the industry.

In a more general way, though, in staking its international strategy on entry into developed industrial markets and on acquisitions rather than new start-ups, Vivendi is very much in line with the whole history of foreign direct investment in manufacturing and services. As we know, most foreign direct investment flows among advanced economies rather than from the developed to the developing world. And most of this happens through the acquisition of existing firms rather than through greenfield investment (Dicken, 1998a; 2000; Schoenberger, 1997; Shatz and Venables, 2000). So from this perspective, the internationalization of environmental management looks a lot like everything that preceded it. Perhaps even more so, given that so much of the investment is place-bound; the operations must be physically in the market – so market size and income distribution will be strong determinants of investment.

The fourth aspect of Vivendi's operations that is especially interesting is the emphasis on consolidation and integration of the industry as the guiding strategic mantra. In this case, the integration is seen to proceed simultaneously on the vertical (waste collection to landfill closure) and the horizontal (water, wastewater, solid waste) axes. Consolidation and integration are also normal and recurring features of industrial history. Sometimes the strategy works for greater or lesser periods of time, sometimes it doesn't, for technological or organizational or regulatory reasons. JP Morgan could consolidate railroads, but he wasn't able to consolidate marine transport. The 'inevitable' fusion of communications and computer technologies into one large business has been predicted since the dawn of the computer age and many companies have lost a lot of money betting on its imminent arrival, yet it is still not quite here.

What are the prospects in environmental management? Does the mantra have illocutionary power: if the largest companies say it enough, will it come true? If Vivendi believes this is possible and desirable, and bets enough money on it, will that make the vision materialize, through the company's own actions and through setting an example that other major players will follow?

We can't know the answers, of course, but here are some things I would think about. First, the strategy of consolidation concerns, in the first instance, the advanced industrial economies. This is a very internationalized game, but not a global one yet, or for a very long time. The 'global' markets, though in many ways attractive, are too undeveloped and physically and institutionally fragmentary to provide a platform for consolidation. Participation in them is likely to be also fragmentary and locally opportunistic.

Second, there are features of the environmental management business that make it very unlike consolidating industries in the more distant or even recent past such as many branches of manufacturing or business services. This holds true even in the most advanced markets, those that are the best targets for both consolidation and integration. One of these is how physically implanted and extensive operations must be. You have to be on the ground, you have to cover a lot of ground, and you have to stay put over the long run. This is a very peculiar set of constraints. I don't know that it puts a strategy of meaningful consolida-tion and integration ultimately out of reach except on a localized or regionalized basis within the developed markets or, alternatively, within certain product cate-gories that are not so physically grounded (e.g. filtration equipment), but it is something to think about.

The business also exists within a geographically and operationally diverse regulatory and institutional environment, which means that quite different technologies and systems designs may be appropriate or mandated in different locations. Different kinds of regulations and institutional overseers are involved in water, wastewater and solid waste. What exactly is Vivendi able to bring into foreign markets that will give it a competitive edge there; what 'ownership advantages' does it have that allows it to trump the advantage of being born in the market? Vivendi France, for example, may well have demonstrably superior water technology than anything currently in use in the US – technology that provides cleaner water more cheaply. But if, for regulatory reasons, that technology is ruled out of court in the US, then the ownership advantage is lost.

There are in fact some things the company might have that would make the goal of international consolidation and integration plausible, if not readily achievable. Cheaper access to capital than domestic competitors might be one. Better management and systems engineering capabilities might be another. For complex systems, this might afford a technical edge that is not embodied in the physical plant. There might be economies of scale and scope across the business (i.e. not tied to operations in specific locales) that come into play. It is hard to know, though. The pace of transformation of the business in Western Europe and the US has been so fast that there has not been time to prove the theory behind it. Vivendi Water got suddenly big in the US basically through two investments, in US Filter and in PSC. Both these companies got big in the first place by consolidating smaller operations, but this has also been fairly recent. US Filter, the biggest Vivendi subsidiary in the US, had only a ten-year track record by the time it was acquired. Vivendi got suddenly big in the international waste management business partly by acquiring assets of Waste Management, Inc., which ran into trouble precisely in the course of implementing an aggressive strategy of international consolidation and integration.

The question of the payoffs to horizontal integration is even more difficult to answer. Does a strong position in water supply in fact give you a significant leg up in the solid waste business? Perhaps in industrial markets, but the minute you need to deal with separate bureaucracies in a given municipality, it seems a safe bet that synergy goes out the window.

If Enron and Waste Management, Inc. are any guide, it is not yet evident that there are significant benefits from integration across water supply, wastewater treatment and solid waste management.

In this brief snapshot, we see a firm making large, aggressive bets about the international and organizational future of a set of businesses concerned with environmental inputs and the management of wastes. On both axes we see a pattern that might be described as highly concentrated extension. Geog-raphically, we see a dramatic commitment to internationalization, implemented over an extremely compressed time frame, but firmly oriented to the largest, most advanced and richest industrial markets in the EU and the US. Organizationally, we see a wholesale commitment to consolidation and integration as the key to achieving a dominant and high-yielding position in the industry. These commitments, in turn, frame a range of other processes and considerations having to do with technology development and transfer, mergers and acquisitions, and the like.

THE LARGER PICTURE

To date, the internationalization of the environmental management industry seems to be a rather straightforward recapitulation of the earlier history of foreign direct investment in manufacturing and services. It comes from the same places, it goes to the same places, and it occurs largely through acquisitions rather than greenfield investment. But there are some distinctive features that bear watching, and questions concerning the timing, geography and magnitude of this wave of investment that lead to some broader aspects of the present evolution of the global economy.

One of the distinctive features, as noted earlier, is the peculiar place-boundedness of the investment. As Peter Dicken (2000; in this volume) has reminded us, there is every reason to believe that even the most internat-ionalized transnational corporations remain strongly marked by their place of origin. At the same time, their various offshoots, no matter where located, must in turn be marked by the conditions of life in different areas of the world (see also Schoenberger, 1999; Gertler, in this volume). But the environmental management industry is a key part of the production and maintenance of the built environment – a physically embedded resource system that supports production, circulation and consumption (Harvey, 1982). The complicated social and institutional realities of place are amplified by their intersection with the physical. The industry's output is produced in and through a physical environment shaped by a history of social and economic processes and regulated by a particular institutional nexus. This particular form of international invest-ment implies an extraordinarily grounded encounter between the global and the local and perhaps helps us to understand that relationship in a somewhat different light.

A second distinction concerns the particular way this wave of international investment is related to the question of the national state. Again, as Peter Dicken (1994; 1998a) has taught us, the global appears not to have vitiated the power of the nation-state, as so widely anticipated, although it has perhaps altered its scope and orientation (see also Peck, 2000). The curious thing about this industry (and related investments in urban and industrial infrastructure such as highways) is that it has entered into a sphere that was long considered the natural purview of the state, which either produced the service itself on the basis of tax and perhaps user revenues, or closely monitored and regulated private monopoly providers as to quality and extent of the service offered and appropriate rates of profit.

The centrality of the state seemed to be guaranteed by problems of pricing on the one hand, and, on the other, the extraordinarily long turnover times on the capital invested which impeded the course of accumulation. Or, to put this another way, the investments were *necessary* to private accumulation in other spheres, but were not themselves easily adapted to private capitalist production and the normal rhythms of capital accumulation (See Harvey, 2001 [1975]; 1982).

A whole concatenation of economic, political and ideological movements seem to have produced an historical moment in which it was arguably both more possible and more necessary to open up this sphere of activity to private investment and accumulation. The greater *possibility* economically is plausibly tied in part to innovations in financing that, although risky (viz. Enron), did ameliorate some of the problems traditionally associated with such long-lived, geographically fixed investment. The political and ideological opening was tied to the rise of a particular form of conservative or 'neoliberal' politics, notably in the UK and the US and propagated by them through major supranational institutions such as the IMF and the World Bank (see Tickell and Peck, in this volume). This political programme was particularly vehement about seeing the state's role as primarily about supporting private accumulation and was particularly unconcerned about the social exclusions that might attend the reduction or elimination of the state's traditional role as a provider or regulator of necessary environmental, infrastructural and social services.

The greater *necessity* on the other hand, arises from what was happening elsewhere in the global economy. If we take as a starting point the advent of the Fordist crisis in the late 1960s–early 1970s, we can discern over the subsequent thirty years a strikingly compressed series of international booms and busts as surplus capital sought outlets for productive investment. These hoped-for 'spatial fixes' at different times saw vast capital inflows into 'world city' property markets (New York, London, Tokyo), developing regions (Latin America, Asia), and the former socialist bloc before wheeling back into the virtual space of the dot.com bubble (Harvey, 1982). All ended in glut, devaluation, and frantic bailouts. The violence of this sequence and its persistence over three decades, despite the sheer amount of value that was vaporized in the recurring collapses, hints at how desperate and huge the need was to find ways to employ productively the surplus capital sloshing around in the global economy.

In this context, a general move to open up the erstwhile 'natural' state arena to private investment makes a lot of sense. Privatizing formerly publicly underwritten elements of the built environment is a way of creating a new space to be colonized by normal capitalist processes. Moreover, to the degree that state assets are transferred at a heavy discount into private hands, this new space is at least temporarily protected from the weightiest rigours of capitalist competition. The result is the diversion of at least some of this flood of excess capital into long-run, fixed investment on the surface of the earth (e.g. public transport) and tightly enmeshed with the physical environment (e.g. water, wastewater, solid waste). In this way, both the physicality of environmental management and its peculiar relationship to the state enter into play here in terms of the questions of timing, geography and magnitude signalled above.

We might be witnessing the emergence of a new kind of spatial fix and its entanglement with a particular version of an 'institutional fix' to some of the pressures of the current period. As Peck (2000) observes, an institutional fix is part of a qualitative reorganization of the mode of social regulation and a *rearticulation* of the state with the economy (rather than its absolute withdrawal from it) as a response to and reconstitution of economic pressures and instabilities. The state offers up part of its own domain as a new institutional space for colonization by private capital at the same time that it helps guide this investment into the space of the physical environment. This may be seen as an innovative substitution for more traditional forms of spatial fix involving, e.g. access to new geographic territories (Harvey, 1982).

Will this diversion of private investment into the state sector and the environment work, and for whom, are the obvious questions with no obvious answers. Quite plausibly, it will work extremely well for some firms and not for others. Some kinds of workers in these operations will benefit, although it seems a safe bet that many more will find their conditions of work eroded if not terminated outright as they shift from the state to the private sector (see Peck, 1996). Similarly, real benefits may accrue to some consumers while others face tremendous costs (e.g. British rail passengers or Californian energy consumers).

More broadly, this combined spatial/institutional fix may indeed help to relieve some of the pressure of excess capital for some time, but whether that could be enough to stabilize the global system for any significant period is open to question. Here the persistent concentration of investment in the advanced industrial economies signals that the 'global' market is insufficiently developed by a huge margin to absorb all that needs to be absorbed. Indeed, if the global market had been developed enough to absorb vastly larger and more spatially extensive waves of investment in the 'usual' activities (manufacturing, etc.) – which it quite plainly has not given their continued overwhelming concentration in a handful of countries – then the need for opening up the state/environmental arenas would not have been so pressing in the first place.

The environment would seem to afford almost limitless scope for investment in clean-up and protection. In effect, it could be a most effective physical sponge for surplus capital – but that could only be in a world in which investment was keyed to need and not to effective demand.

ACKNOWLEDGEMENT

My very great thanks to my colleague, John Boland, for his help and advice. My Thanks also to Jamie Peck and Henry Yeung for their comments and suggestions. Most of all, thanks to Peter Dicken for many years of inspiration and support.

NOTES

1   RWE has more recently made a $3.3 billion bid for Innogy, the UK's biggest electricity supplier. Indeed, continental power suppliers have been particularly active in the UK, which is Europe's most liberalized energy market. Electricite de France acquired London Electricity in 1998 as well as part of SWEB. Meanwhile, Eon, Germany's largest power supply group and RWE's largest competitor, is in the process of acquiring Powergen for £10 billion (*Financial Times*, 2002a).

2   The 63 per cent ownership share is as of this writing in spring, 2002. It is anticipated that Vivendi Universal will sell off a further share of Vivendi Environnement, dropping its holding to below 50 per cent, in order to reduce the parent company's debt burden. French political concerns about foreign ownership of the subsidiary are expected to complicate any sell-off, however, leading to expectations that the largest acquirer of shares will be Electricite de France, the state electricity monopoly (*Financial Times*, 2002b).

3   It may also be the case that access to capital has been facilitated by innovative debt financing. One wouldn't have to go quite so far as Enron and Waste Management to attract capital to such an apparently explosive growth sector. What is certainly the case is that the parent conglomerate's debt burden (here I mean Vivendi Universal) is now considered excessive, although this is also related to acquisitions in the media sector.

PART TWO

PLACING GLOBAL KNOWLEDGE

Chapter 6

THE SPATIAL LIFE OF THINGS:
THE REAL WORLD OF PRACTICE WITHIN THE GLOBAL FIRM

Meric S. Gertler

> ... contrary to much received wisdom, place and geography still matter
> fundamentally in the ways in which firms are produced and in how they
> behave. Despite the unquestioned geographical transformations of the world
> economy, driven at least in part by the expansionary activities of transnational
> corporations, we are not witnessing the convergence of business-organizational
> forms towards a single 'placeless' type. Organizational diversity, related at least
> in part to the place-specific contexts in which firms evolve, continues to be the
> norm. (Dicken, 2000: 287)

INTRODUCTION

As Peter Dicken (1998a; in this volume) has frequently reminded us, the global firm
is one of the dominant agents redrawing the map of the world's economy. This
influence is felt most profoundly through the international processes of foreign direct
investment as well as mergers and acquisitions (see Schoenberger, in this volume).
Both of these global flows create the potential for the transposition of production
technologies and industrial practices from one regional or national economic space to
another. Indeed, it is commonplace these days to see these processes as leading to
convergence in national and regional industrial practices over time, with the
transnational corporation (TNC) acting to diffuse a wide array of 'best practices'
throughout its global operations (Gertler, 2001).

However, this interpretation is not uncontested: others view the same process and
see something entirely different – a world in which the global flows of technology and
practice both produce and are produced by geographical difference. In particular,
contemporary accounts within the international political economy literature vigorously
debate the role (or lack thereof) played by nation-states in mediating this process.

Meanwhile, economic geographers have argued that the sub-national region-state has become more, not less, important as a source of institutional influence over the practices and actions of private firms. While some see this newly expanded regional role as arising to fill a void created by the contraction or 'hollowing out' of the nation-state's institutional capacities, others see the same phenomenon as part and parcel of a reinvigorated agenda for public action to stimulate economic and technological development in the private sector. In short, nations and regions both continue to matter.

This chapter identifies key issues in the current debates on these questions, with a particular focus on the specific forces both enabling and constraining the production and deployment of industrial technologies and practices in particular places. In distinction to the vast literature in management studies that conceives of industrial practices as the direct outcome of individual and corporate decision-making, the approach taken in this chapter emphasizes the interaction between individual or corporate actions and the institutional environment within which such actions take place. The chapter proceeds by addressing three central questions. First, what kind of influence do the institutions of the nation-state continue to exert over the practices of private firms? Second, to what extent do transnational firms 'learn' across international boundaries, and how does this affect the continuing survival of distinctive national industrial models? Third, how locally or regionally embedded is the global corporation and where is such embedding most likely to take place? The chapter concludes with a sober reappraisal of the 'real world' of industrial practice within the global firm, one in which local social context – what we might think of as 'the spatial life of things' – matters a good deal more than most accounts have acknowledged thus far.

WHAT REMAINS OF DISTINCTIVE NATIONAL INDUSTRIAL MODELS?

In the past few years, an important and lively debate has taken shape concerning the influence of the nation-state and whether or not it retains the institutional purchase necessary to regulate and govern the national macro-economy. So-called 'gurus' of international business, such as Kenichi Ohmae (1995), have famously argued that the transition to a truly global economy has dramatically undercut the power, influence and importance of the traditional nation-state, with other (sub-national) institutional and spatial forms arising to take its place. Related to this, others have argued that the institutions of national states in Europe and elsewhere are being transformed increasingly to resemble a new Anglo-American norm or standard (Albert, 1993; *The Economist*, 1996).

Critics such as Hirst and Thompson (1996) and the contributors to Cox (1997), Berger and Dore (1998) and Boyer and Hollingsworth (1999) have raised serious concerns about the empirical accuracy of such claims. They argue that nation-states continue to wield many important powers in the economic realm (see also Dicken et al., 1997a; Leyshon, 1997). A growing body of research in geography, economics and international political economy asserts the following thesis. Notwithstanding the undeniably increasing influence of supra-national institutions such as the European Union, the North American Free Trade Agreement and the World Trade Organization, and while acknowledging the growing importance of the region-state and other non-state regional initiatives, the nation-state and its institutional legacy still continue to exert a crucial influence over the practices of firms. However, the manner in which this influence is exerted remains the subject of some debate.

A landmark study by Doremus et al. (1998) performed on behalf of the US Office of Technology Assessment examined the practices of American, German, and Japanese TNCs (both at home and abroad), focusing on specific functions including corporate governance, finance, research and development, and intra-firm investment and trade. The authors conclude, with few caveats or qualifications, that strongly distinctive national models remain evident in the continuing practices of these large firms. They argue that these TNCs are forever shaped by their national origins. Although these firms do respond and react to (or anticipate) changing competitive conditions, the strategy or path they choose to follow is strongly influenced by the national institutional legacy of their home country (for a more succinct summary of these findings, see Pauly and Reich, 1997).

Such conclusions are consistent with the body of work emanating from the national systems of innovation literature. The contributors to Lundvall (1992), Nelson (1993), and Edquist (1997), as well as Pavitt and Patel (1999) and Freeman (1997), assert that distinctive nationally organized constellations of institutions shape firms' innovation practices and longer-term strategic and technological trajectories. Drawing on evolutionary theory in economics (Nelson, 1995), they argue that these sets of institutions, once in place, generate slowly changing, path-dependent processes over time. Moreover, there is no compelling evidence of convergence in national activities for technological accumulation since the 1970s and even some evidence of divergence, implying that national innovation systems have become more (not less) distinctive over time.

A related literature, on national business systems and 'varieties of capitalism', offers a complementary perspective that goes well beyond the study of innovation-generating activities alone to argue that virtually *all firm practices* (day-to-day practices as well as long-term strategies) are strongly influenced or governed (though not wholly determined) by national macro-regulatory institutions and 'market rules' (Maurice et al., 1986; Christopherson, 1999; 2002; Streeck, 1996; Whitley, 1999; Lazonick, 2000; O'Sullivan, 2000; Hall and Soskice, 2001; Lam, 2000). Moreover, firms are quite often not consciously aware of the influence that this larger institutional matrix exerts on their choice of practices. This literature argues that the contours and composition of the national macro-regulatory framework make certain choices easier or more likely, and others less so. Thus, firms' time preference structures and investment payback rules are shaped by the structure of the capital markets in which they principally raise their investment finance (i.e. at home). Corporate strategy is also bounded and constrained by these same national capital market structures and characteristic national systems of corporate governance. Employment practices and relations and training regimes are strongly shaped by nationally distinctive systems of education, training, labour market and industrial relations regulation. These time horizons, labour market practices and industrial relations systems in turn shape technology choice and use.

Even firms' predisposition for cooperation and collaboration with customers, suppliers and competitors is shaped by domestic competition policy, labour market regulation and trading rules (Arrighetti et al., 1997; Lane, 1997).

Notwithstanding the impressive weight and consistency of these findings, it would be misleading to imply complete consensus on the matter of *how* nation-states exert their governing influence over (even transnational) economic actors. In particular, available research on this subject suggests that national institutions of the host (rather than home) economies are most influential in shaping the practices and strategies of transnational firms abroad. Focusing on dimensions of workplace organization such as the distribution of autonomy and power, decision-making with respect to new technologies and the use of teams or other participatory methods, Wever (1995) examines the experiences of American firms operating branches in Germany as well as German firms with branches in the US. In a detailed assessment of these cases, she reports that both sets of firms were frustrated in their attempts to institute the 'home' way of doing things in their foreign branch operations. Hence, over time, the US firms' branch operations in Germany took on an increasingly German style of industrial relations and work practices, despite the parent firms' initial intent. The same is true of the German firms' branch operations in the US. Hence, she claims that when MNCs go abroad, the rules of the host country hold sway.

In my study of the transfer of German practices and technology abroad, I conducted interviews with German advanced machinery manufacturers serving the Canadian and American markets (Gertler, 1996), having also interviewed their customers in Canada. These interviews provided a window on the very different ways in which advanced machinery designed and built in Germany was implemented in North American workplaces compared to German workplaces. The analysis of this information focused on eight specific dimensions of machinery implementation and use in order to provide a structured framework with which to compare industrial practices in North America and Germany (Gertler, 2002a). In two of the fifteen German case studies conducted, the machinery-producing firms were supplying manufacturing systems to their own branch plants in North America that were producing other products under the same corporate umbrella. In essence, these particular examples became case studies of the parent firms' attempts to transpose their distinctively German system of production to a North American workplace. The overwhelming conclusion of this work was that the most fundamental sources of operational difficulties for the North American users (including German-owned user plants) were to be found in the starkly different macro-regulatory environment and institutions regulating labour markets, industrial relations, corporate governance and capital markets in these two 'models' (Gertler, 1999). In other words, the 'host rules' prevailed here too, providing an inhospitable foundation for the implementation of German manufacturing practices in North America, even when the firms attempting this implementation were themselves German-owned.

Similar results are evident in Abo's (1996) cross-national comparison of Japanese-owned 'transplant' operations in the automotive and electronics indust-ries. In seeking to determine the degree of completeness with which the Japanese model of production was transposed to these foreign plants, he found sizeable differences in the manner in which the 'model' was actually applied. The emerging 'transplant geography' documented by him reflects the strong differences in national regulatory frameworks and institutions evident within his sample of 'hybrid factories'. The strongest concordance with the Japanese model was found in transplants in those nations (such as Korea) whose national regulatory frameworks for corporate governance, industrial finance, labour markets and industrial relations most strongly resemble Japan's. Not surprisingly, the weakest concordance was found in American transplant operations.

Finally, Schoenberger's (1999) analysis of US multinationals with Japanese branches lends further support to the argument that rules in the national 'host' setting prevail. Her case study of Xerox corporation reveals that the American parent had ample opportunity to learn from the obviously successful practices of its Japanese subsidiary (a joint venture with Fuji), but failed to do so. Her analysis raises some important questions about the difficulties or barriers to intra-firm learning across major boundaries between distinct national systems (a theme to which I shall return below; see also Amin, in this volume).

The work reviewed above offers many important insights in support of the general proposition that the nation-state (whether 'home' or 'host') is still a primary source of influence over industrial practices. Moreover, this work argues implicitly that *all firms are embedded* within an institutional matrix, whether they realize it or not – what Appadurai (1986: 49) has described as 'the cultural design of capitalism'. This is *not* a process that happens just in those places lucky enough to have abundant trust and local institutional thickness. As a result, as we shall discuss at greater length in the following section, the success with which firms can transpose a distinctive set of practices from one national space to another (i.e. 'learn') where the institutional environment is not as conducive or supportive of such practices will be limited at best.

DO GLOBAL FIRMS LEARN ACROSS INSTITUTIONAL BOUNDARIES?

By definition, inter-organizational learning implies the transfer of knowledge between individual firms, between establishments within the same firm, and/or between firms and other organizations around them such as universities, research centres and service centres. What do we know about the processes by which such transfers take place? Recently, much attention within the social sciences has been directed to the fact that a large component of firms' most valuable practical knowledge is tacit in nature. The principal distinguishing traits of such tacit knowledge are:

1   It is difficult to communicate effectively through written – and sometimes even verbal – form.

2   It often resides in the unconscious realm of knowledge, meaning that individuals who hold such knowledge may remain unaware of 'what they know' or how this knowledge shapes their attitudes and practices.

3   It is context-specific – in other words, one's ability to absorb and share such knowledge depends to a large extent on the degree to which one is embedded within the social context in which such knowledge is produced.

The problem of how to spread tacit knowledge around more widely, once it has been produced and identified as being useful, has received growing attention in recent times. While it is an issue of central importance to economic geographers for obvious reasons (see Amin, Thrift and Malmberg, in this volume), it has recently become a concern of overriding importance within the knowledge management literature as well as within industrial economics and the innovation systems literature, where it has been approached as a problem of how to promote social learning processes. The general consensus emerging from this literature is that, in a world in which access to codified knowledge is becoming ever easier, a firm's ability to produce, access and control tacit knowledge is most important to its competitive success (Nonaka and Takeuchi, 1995; Prusak, 1997; Maskell and Malmberg, 1999).

This issue presents a special problem for 'distributed organizations' in which different units are situated in different locations separated by long distances and the tacit knowledge associated with desirable ('best') practices must be transmitted across regional and national boundaries as well as cultural and other divides. If there is one assertion on which there is widespread agreement, it is that the transmission or diffusion of tacit knowledge is not straightforward. This is principally because successful sharing depends on close and deep interaction between the parties involved. However, as we shall see below, there is considerable disagreement concerning how 'close' should be defined. For example, the learning regions perspective (Morgan, 1997; Cooke and Morgan, 1998; Lundvall and Maskell, 2000; Leamer and Storper, 2001) argues that tacit knowledge does not 'travel' easily. This is because its transmission is best shared through face-to-face interaction between partners who already share some basic similarities: the same language; common 'codes' of communication; shared conventions and norms; personal knowledge of each other based on a past history of successful collaboration or informal inter-action. These commonalities are said to serve the vital purpose of building trust between partners, which in turn facilitates the local flow of tacit knowledge between partners.

The economic geography associated with this perspective is well known: since spatial proximity is key to the effective production and transmission/sharing of tacit knowledge, this reinforces the importance of innovative clusters, districts and regions. Moreover, as Maskell and Malmberg (1999) point out, these regions also benefit from the presence of localized capabilities and intangible assets which further strengthen their centripetal pull. Many of these are social assets, i.e. they exist between rather than within firms. These assets also include the region's unique institutional endowment, which can act to support and reinforce local advantage. Because such assets exhibit strong tendencies of path-dependent development, they may prove to be very difficult to emulate by would-be imitators in other regions, thereby preserving the initial advantage of 'first mover' regions.

Another recent body of work argues that while organizations may be able to produce tacit knowledge effectively, the active dissemination of this knowledge more widely within and between organizations constitutes a serious – though not insurmountable – management challenge for the firm. This problem has become the focus of a huge effort by firms (especially large ones), and has come to be recognized – even by those originally promoting the idea of a 'knowledge-creating company' – as a very significant obstacle to greater innovativeness (Ichijo et al., 1998; von Krogh et al., 2000). This literature sets about to document some of the creative ways in which firms have responded to this situation, emphasizing the key role of knowledge enablers – that is, 'knowledge activists' who aim to span boundaries within the large firm, acting as agents for the diffusion of tacit knowledge – normally with at least partial codification in the process of transmission.

The boundary-spanning strategies of these knowledge activists make heavy use of story-telling as a key mode of tacit knowledge transfer. But even this can only work when supported by direct, face-to-face interaction and communication between people. For this reason, another key element of a knowledge-enabling strategy is the circulation of key personnel between head office and branch locations (or between different branches) around the globe (see Coe et al., in this volume). Hence, according to this view, although the production of tacit knowledge remains strongly localized, the possibilities for its dissemination – once produced – may lead to large spread effects within multi-divisional and multi-locational firms. There is also at least the potential for wider diffusion of this knowledge outside the firm, but this requires the firm to devote considerable resources toward the dissemination of tacitly inflected best practices.

Von Krogh et al. (2000) argue that 'microcommunities of knowledge' play a key role to ensure the success of this tacit knowledge circulation within large organizations. These are small groups of five to seven people who are strongly bound together through common work histories and who rely crucially on face-to-face interaction for their success. 'Geography' – that is, both physical separation and local cultural differences – is acknowledged as making all of this more difficult and challenging. While emphasizing the importance of a common or shared 'social context' in facilitating the flow of tacit knowledge, they view the creation and shaping of this context as primarily within the purview of the firm, admitting no explicit role for the institutions in which economic action is situated. Furthermore, although their detailed case studies document firms' efforts to facilitate the transmission of tacit technological knowledge between culturally (as well as physically) distant sites, they provide no real insights into how 'local culture' is produced.

Closely related to the 'microcommunities of knowledge' concept, another recent literature emphasizes the central role of 'communities of practice' as key entities driving the firm's knowledge-processing activities. This literature argues that routines and established practices shaped by organizations (or subsets of people within organizations) promote the production and sharing of tacit knowledge (Wenger, 1998; Wenger and Snyder, 2000). Communities of practice are defined as groups of workers informally bound together by shared experience, expertise and commitment to a joint enterprise. These communities normally self-organize for the purpose of solving practical problems facing the larger organization and, in the process, they produce innovations (both product and process). The commonalities shared by members of the community facilitate the identification, joint production and sharing of tacit knowledge through collaborative problem-solving assisted by story-telling and other narrative devices for circulating tacit knowledge.

Thus, according to this approach, organizational or relational proximity and occupational affinity are more important than geographical proximity in supporting the easy flow of tacit knowledge (Amin, 2000; in this volume; Amin and Cohendet, 2000). The resulting geography is distinctly different from that which is envisioned in the learning region approach, and also differs subtly but significantly from the 'knowledge enablers' approach reviewed above. In this view, the joint production and diffusion/transmission of tacit knowledge across intra-organizational bound-aries is possible, so long as it is mediated within these communities. Moreover, because communities of practice may extend outside the single firm to include customers or suppliers, tacit knowledge can also flow across the boundaries of individual organizations. Most importantly, the communities of practice literature confidently asserts that tacit knowledge may also flow across regional and national boundaries if organizational or 'virtual community' proximity is strong enough – what might be thought of as the 'de-territorialization of closeness' (Bunnell and Coe, 2001). In other words, learning (and the sharing of tacit knowledge) need *not* be subject to the 'friction of distance', if relational proximity is present.

A notable departure from this position is the work of Brown and Duguid (2000) who express far less optimism about the potential for long-distance collaboration through communities of practice. While they acknowledge the important role of communities of practice in promoting social knowledge production and circulation, they conclude that face-to-face interaction is an essential and non-substitutable ingredient for successful collaboration leading to innovation within such communities. They argue that tacit knowledge cannot be assumed to circulate freely just because the technology to support its circulation is available. In their view, the narratives and social ties so crucial to the flow of knowledge within communities of practice are deeply embedded within the social systems in which they arise.

The communities of practice approach offers useful reminders of the im-portance of relationships and the strength of deeper, underlying similarities rather than geographical proximity *per se* in determining the effectiveness of knowledge-sharing between economic actors. However, this argument leaves unanswered one very important question: what forces shape or define this 'relational proximity', enabling it to transcend physical, cultural and institutional divides? How are shared understandings produced? As I have argued recently (Gertler, 2002b), we cannot sort out the geography of tacit knowledge and the possibilities for international learning by the global firm without inquiring more systematically into the fundamental nature of 'culture' and the institutional underpinnings of economic activity. If we accept Karl Polanyi's (1944) basic precept that markets and the behaviour of economic actors are socially constructed, embedded and governed, then this suggests that the ability of individual workers or firms to produce and share tacit knowledge depends on much more than spatial proximity, cultural commonality or organizational affinity. In particular, it depends on institutional proximity – that is, the shared norms, conventions, values, expectations and routines arising from commonly experienced frameworks of institutions, as discussed in the previous section of this chapter.

Indeed, recent case studies of attempted knowledge transfers within and between global firms indicate that this form of proximity or affinity often overrides organizational or relational proximity when the organization in question extends geographically across institutional divides (Schoenberger, 1999; Gertler, 1999; 2002a). Therefore, we can more productively interpret the origins of routines, characteristic practices and 'settled habits of thought' as arising from concrete institutional origins: while corporate agency and the distinctive 'culture' of the firm undoubtedly play a major role, they do not exist within a vacuum (Glasmeier, 2001) and, contrary to the underlying premise of much of the knowledge management literature, managers do not fully (or even largely) shape their own destiny. They operate within a possibility set that is constrained by larger forces – particularly the institutional and regulatory frameworks at the national and regional scales (Lam, 1998; Lam and Lundvall, 2000). Moreover, such institutional influences are subtle but pervasive: indeed, often so subtle that firms and individuals are not even conscious of the impact they exert over their own choices, practices, attitudes, values and expectations.

HOW LOCALLY EMBEDDED IS THE GLOBAL CORPORATION AND WHERE?

Recapping the story thus far, it seems apparent that the overall strategic direction and organization of the global firm – with a few exceptions – remains strongly influenced by the institutional framework prevailing in its country of origin. Nevertheless, the nature of its operations at branches outside the home base are likely to be more strongly shaped by the institutional frameworks of the host country. Does this mean, however, that the firm is embedded to the same degree in each of its sites of operation? Put in a slightly different way, when firms invest in operations abroad, is the 'behaviour' of these foreign-based establishments in their host environment indistinguishable from that of indigenous firms? This question has taken on added significance in recent years, particularly amongst those who have been studying the changing international organization of the innovation process itself – that is, research and development, and related downstream product and process modifications. Not surprisingly, one finds a lively debate and a diversity of views on the nature of emerging trends and patterns.

Predictably, there are those who argue that the rise of information technology and global telecommunication networks has provided firms with the logistical capability or potential to organize and coordinate their R&D and their acquisition of technical knowledge on a global basis. Ostry and Nelson (1995: 24) have coined the term 'techno-globalism' to convey their view that transnational corporations are establishing their research activities in key R&D centres distributed across a number of countries and building alliances with university research centres as well as with other firms offering complementary knowledge and skills. Some recent studies (Kuemmerle, 1999; Granstrand, 1999; Pearce, 1999; Florida, 1997) document a steady increase in the trend towards decentralizing R&D, with firms locating research facilities abroad to augment their existing knowledge base and technological capabilities or to exploit their existing capabilities through product development for local markets. Critical to this is the growth in industry-university collaborations among foreign TNCs, as illustrated by the growing investment of foreign firms in leading US-based research institutes in order to benefit from the intellectual output of the US research system (Florida, 1997).

Analysing patent data, Cantwell (1999) similarly maintains that globalization and national specialization are complementary parts of the process and not conflicting trends. The trend towards organizing R&D on a global basis is founded on the desire to tap into the locally specific and differentiated stream of innovation in each centre (see also Cantwell and Janne, 1999). In this vision, which might be thought of as describing a 'post-national' innovation system, one would expect differences in innovativeness between domestic and foreign-owned firms to be small or absent altogether, as foreign-owned firms locate these activities overseas to take advantage of location-specific technological capabilities.

However, a considerable amount of recent evidence supports the contrary view: that despite the increasingly global nature of technological activities, national differences among the leading industrial countries remain significant and the specific character of the home base is crucial to the innovativeness of domestic firms. Patel and Pavitt (1997) conclude that the evidence on R&D and US patenting activities by the world's leading TNCs shows no evidence of convergence in national activities for technological accumulation since the 1970s and even some evidence of divergence (see also Pavitt and Patel, 1999; Patel and Vega, 1999). Their findings, which are also consistent with the work of Pauly and Reich (1997) and the larger literature on national innovation systems, indicate that technology-generating activities of TNCs remain the most domesticated of all corporate activities. They find only limited evidence to support the view that firms go abroad to compensate for their core weaknesses at home. The dominant factor for locating R&D activities abroad in core industries continues to be adapting products and processes for foreign markets and providing technical support to overseas manufacturing facilities.

Under this scenario, the foreign operations of such firms may support innovative activity, but this is more likely to be confined to the customization of existing technologies to suit the tastes or unique conditions of local markets. Linkages to science-intensive local/regional universities and public research labs would logically be of less importance for such foreign-owned firms, especially when compared to the practices of domestically based counterparts. For indigenous firms, linkages to local sources of innovative ideas, whether in the innovation-supporting infrastructure of universities and research laboratories, in the capabilities and demands of customer and supplier firms up *and* down the supply chain, or in the complementary knowledge assets of competitors, ought to constitute the very essence of their 'home base'. Overall, the evidence provided by the work of Pavitt, Patel and colleagues suggests that the role of the home country (or 'home base') and their individual policies is *not* reduced as a result of globalization (see also Dicken, in this volume). Notwithstanding the globalization of markets and production, there remains a compelling reason why companies continue to concentrate a high proportion of their technological activities in their home base.

> The development and commercialization of major innovations requires the
> mobilization of a variety of often tacit (person-embodied) skills and involves
> high uncertainties. Both are best handled through intense and frequent
> personal communications and rapid decision making – in other words, through
> geographic concentration. (Patel and Pavitt, 1994: 773)

This conclusion is in some ways consistent with the insights offered by the recent literature on the relationship between globalization, the rise of the knowledge-based economy and the role of regions and localities. As noted above, the argument here is that spatial proximity facilitates frequent, close and (most commonly) face-to-face learning-through-interaction (Lundvall and Johnson, 1994). Second, as noted in the earlier discussion of the learning region thesis, firms clustered in the same region often share a common regional culture that can act to facilitate the process of social learning, especially when much of the most important knowledge transmitted between parties in the innovation process is tacit rather than codified. Finally, this interaction-facilitating common language or code of communication is further supported by the creation of regional institutions that help to produce and reinforce a set of rules and conventions governing local firms' behaviour and inter-firm interaction.

In the past few years, the constellation of institutions at the regional level contributing to the innovation process has come to be known as the regional innovation system (Braczyk et al., 1998) in a manner analogous to the concept of national innovation systems. The basic argument here is that this set of institutions, both public and private, produces pervasive and systemic effects that encourage firms within the region to adopt common norms, expectations, values, attitudes and practices – in short, a common culture of innovation that is reinforced by the social learning processes outlined above. The list of institutions most frequently implicated in this type of analysis includes not only the usual R&D infrastructure (universities, technical colleges, public and private labs), but also industry-specific service centres for technology transfer and market analysis, local training councils, producers' associations, chambers of commerce and suppliers' clubs, all of which provide opportunities for social learning-through-interaction. Returning to our discussion of the global economy, the learning region literature asserts that even global firms need to draw their innovative sustenance from the social production system surrounding them. Hence, the model that may best describe this relationship is one of 'local nodes in global networks' (Amin and Thrift, 1992).

However, there is a somewhat troubling ambiguity here. Namely, do TNCs embed themselves in these locally grounded learning dynamics (i.e. learning regions) at home, abroad or both? Unfortunately, the learning region literature is quite indeterminate on this question and no clear picture emerges. Therefore, the important empirical question arising from this work remains: do firms seek to embed themselves in such interaction-rich learning regions, no matter where they might be found, or will they keep the bulk of their innovation-generating activity in those 'home' regions whence they originated? This is clearly a question of paramount importance for regions where levels of foreign ownership in manufa-cturing and services are high.

What does the recent evidence tell us about the extent to which such relationships have in fact begun to emerge? The economy of Ontario, Canada's largest and most prosperous province, makes for a fascinating test case in which to examine these questions. Foreign-owned firms have traditionally dominated the province's manufacturing economy and still account for a very large proportion of manufacturing output (nationally, this proportion stands at more than 55 per cent) (OECD, 1998). So, have these foreign-owned firms begun to behave more like indigenous firms in terms of their innovation-producing activities, consistent with the idea of a post-national system of innovation? Have they embedded themselves more deeply in relations with suppliers, customers, competitors, universities, research centres and other innovation partners within an Ontario-based learning region?

The consensus view emerging from recent empirical work is that foreign-owned manufacturing establishments have *not* overwhelmingly rushed to embrace these new possibilities (Gertler and DiGiovanna, 1997; Britton, 1999; Gertler et al., 2000). To the extent that they have formed collaborative relationships with other entities (whether these be firms or institutions), their partners tend to be found outside Ontario and Canada. For American-owned firms, these links are strongest with innovation partners in various regions of the US, following paths that are well-established through existing corporate relationships. Set alongside further evidence documenting the 'hollowing out' of foreign-owned corporations' Canadian headquarter operations (Arthurs, 2000), this evidence would suggest that in the post-NAFTA era, foreign capital is no more deeply embedded in Canada's learning regions than it was before. On the other hand, Gertler et al. (2000) find that Canadian-owned firms in their sample have begun to develop closer, collaborative ties with local customers, suppliers and innovation-supporting institutions (universities, research labs, technology transfer centres) in their home regions. They conclude that, despite the pervasive rhetoric about the global economy, nationality of ownership does still influence the behaviour and practices of private businesses.

If such results are supported by subsequent studies, their implications are significant. The upshot of this finding is that, while the rules and institutions of 'host' economies do tend to prevail – at least when it comes to the kinds of in-house practices around workplace organization, employment relations and labour market practices employed in the branches of foreign-owned firms – this does *not* mean that foreign firms behave exactly like indigenous firms. In particular, the tendency for TNCs to embed themselves in local learning relationships abroad seems still to be quite limited at best. In other words, foreign-owned firms may adopt the same labour market and training practices as their local counterparts, but will not generally adopt the same kinds of relationships crossing the boundary of the firm – a finding of considerable importance.

THE SPATIAL LIFE OF THINGS

In the discussion above, we have argued that the enduring, path-dependent institutions of the nation-state retain far greater influence over the decisions and practices of corporate actors than the current prevailing wisdom would allow. This influence is systemic, pervasive and usually unobtrusive. It shapes – though does not wholly determine – a wide range of industrial practices and knowledge in conscious and unconscious ways, and renders much of this knowledge to be so context-specific and embedded that it cannot be readily transmitted across institutional boundaries without significant loss of meaning and value.

In practical terms, this means that the global firm is not quite so global as it appears to be (see also Dicken, in this volume). In particular, its ability to impose a uniformity of practices dictated by head office across the far-flung outposts of its international operations is in fact significantly limited. So too is the ability of the global firm to 'learn' from its experiences abroad, by transposing successful practices from one institutional environment to another. As a result, the commonplace corporate discourse about implementation of global 'best practices' seems to be implausible at best. And while the overall strategic direction and organization of the global firm are likely to remain strongly influenced by the institutional framework prevailing in its country of origin, the operations at its overseas branches are likely to be more strongly shaped by the institutional contours of the host setting. Nevertheless, when it comes to the performance of core functions such as research and development, these still remain overwhelmingly concentrated at the 'home base' for most firms, industries and countries. Similarly, it is now clear that the rhetoric surrounding the 'convergence narrative' is substantially overblown. In reality, the institutional impediments to global convergence around one (Anglo-American) set of practices are significant and enduring. This is *not* to say, of course, that nation-states and the institutional models on which their economies are based, are immune to change – indeed, far from it. Nor is it to deny that some of the most important forces leading to domestic institutional change are international in origin.

This seems an appropriate point at which to return to the words of Peter Dicken quoted at the beginning of this chapter. The arguments presented above have tried to show how indeed 'place and geography still matter fundamentally in the ways in which firms are produced and in how they behave' (Dicken, 1998a: 287). If the general tenor of these arguments is correct, then they suggest that there remains considerable room for national (and, by implication, local and regional) politics to fashion distinctive strategies for economic development. Indeed, some have even gone so far as to suggest that the role of domestic political and institutional choices has become more, rather than less, important in recent times as nations and regions adopt the objective of differentiating themselves more sharply from their competition (Freeman, 1997; Pavitt and Patel, 1999). If true, then a paradoxical outcome of globalization processes is the renewal of local and national political empowerment. At a time in which many decry the erosion of economic sovereignty of the nation-state, this analysis is salutary. Moreover, it suggests that the conscious choices and decisions made by democratically elected governments retain the ability to influence regional and national economic outcomes. In short, politics – and geography – still matter.

Chapter 7

SPACES OF CORPORATE LEARNING

Ash Amin

INTRODUCTION

A striking aspect of Peter Dicken's work on transnational corporations (TNCs) is its elegantly light touch. Yes, there is clear recognition of the power and global reach of TNCs, their decisive influence on regions and nations, and their centrality in the process of uneven development. But, the giant is also seen to dance on pinched toes, to many tunes and in different directions, responsive to context and circumstance, forcing and forced. This is most evident in his analysis of the mutual constitution of firms and territories. Against claims that the rise of the global corporation represents the annihilation of geography and the erosion of spatial difference, he has consistently argued that *'places produce firms* while *firms produce places'* (Dicken, 2000: 276, original italics; see Dicken, in this volume). For him, the form and behaviour of TNCs continue to display the influence of the national business systems, institutional arrangements and technological regimes in which they developed. They shape TNC relation-ships with the 'home base' over the long run, as well as the nature of internat-ional investments, e.g. a modest home base for innovation encourages R&D-based internationalization. In turn, the 'economies of *places* reflect the ways in which they are 'inserted' into the organizational *spaces* of TNCs either directly, as the geographical locus of particular functions, or indirectly through customer-supplier relationships with other (local) firms' (Dicken, 2000: 283, original italics). The outcome is a corporatized global space made up of varying corporate geographies with varying locational interactions.

I seek a similar lightness in this chapter on the spatiality of tacit learning in distributed organizations. The rise of the knowledge-based approach to the firm has opened promising avenues of research into the geography and sociology of organization. In this approach the firm is conceived as 'a processor of know-ledge' (Fransman, 1994): a locus for setting up, selecting, using and developing knowledge. It differs from the traditional contract-based approaches to the firm – transaction costs theory in particular – which consider the firm as a 'processor of information', a locus for maximizing the efficient allocation of resources. It recognizes that competences – sets of human and technological capabilities – constitute the main sources of the competitiveness of a firm. In doing so, it brings to the fore in the theory of the firm the problem of how such knowledge is generated, maintained and transmitted. Corporate dynamism, even in the all-powerful TNC, is predicated upon an architecture and dynamic of learning that enables firms both to exploit their core business strengths and explore new opportunities.

The knowledge-based approach, through its moorings in evolutionary economics, cognitive psychology, and organizational science, has made enormous gains in recognizing learning and innovation as an uncertain, bounded, and relational science. The key propositions, helpfully summarized by Morgan (2001: 6), are that:

> innovation is in general a groping, uncertain, cumulative and path dependent process ...; agents, be they individuals, firms or states, are subject to bounded rationality ...; tacit capabilities are localized and embedded in individuals and organizational routines ...; firms, and other organizations, display an awesome range of capabilities and cognitive frameworks ...; knowledge is spatially 'sticky' and that tacit knowledge, despite the growth of knowledge management tools, is not easily communicated other than through personal interaction in a context of shared experiences.

These propositions help to explain why innovation – the foundation of corporate survival and adaptation – is differentially and unevenly achieved, why corpor-ations have to organize for and work at learning, and why there are clear limits to what can be achieved. The knowledge-based approach, in a strong sense, highlights the competitive struggle of firms.

My aim in this chapter is to edge the knowledge-based perspective towards an acknowledgement of weakly cognitive practices, or more accurately, knowing through the habits of everyday interaction (see also Thrift, in this volume). I wish to draw on an anthropological sensibility – largely absent in evolutionary and cognitive thought – on the generation of meaning and novelty through the acts of distributed 'communities of practice'. These communities act as performative spaces for both routine and strategic learning based on sociality, mobilizing tacit and codified knowledge in an indivisible continuum. In making this step, there are three points I wish to make. First, an emphasis on learning through social practices challenges evolutionary and cognitive theory, which tends to reserve specific types of competence or cognition for specific types of learning (e.g. procedural rationality for intra-paradigm innovation and reflexive rationality for radical innovation, or tacit knowledge for non-imitable learning and codified knowledge for standardized learning). Second, the emphasis on sociality suggests that the tool-bag for encouraging and managing learning in large organizations requires instruments that are different from those traditionally assumed by the competence-based approach.

Third, *contra* current fashion in geography that the interactive character of learning places a premium on face-to-face contact, and through this, localized networks (see Gertler and Malmberg, in this volume), I wish to suggest that relational proximity is also possible in distanciated networks, through mobility and a series of other technologies of contact and translation.

ON ORGANIZATIONAL LEARNING

Until a decade ago, the literature on TNCs was not centrally concerned with knowledge as the source of competitive advantage. TNC power was explained in terms of scale advantages, monopoly practices, transfer pricing, market access and cost savings from internationalization, deals with governments, and so on. If knowledge was brought into the equation, this was primarily in terms of the economics of technological innovation in new products and processes, the organization of R&D within and across the boundaries of the firm, and the acquisition of new skills and competences. Innovation was seen to flow from science and technology, learning was defined in terms of the acquisition of the relevant know-how, and organizing for technological change and transfer was seen as the associated management challenge. But here too, discussion was restricted to firms operating in high technology or research intensive industries.

The picture is very different now, not only in terms of what counts as knowledge, but also its significance for corporate survival. For example, in the last few years, the influential *World Investment Report* produced annually by the United Nations Conference on Trade And Development (UNCTAD) has increasingly come to emphasize the centrality of knowledge for TNC compe-titiveness. This shift comes with the acknowledgement that the terms of competition have changed, for example, towards: 'the increasing importance of all forms of intellectual capital in both the asset-creating and asset-exploiting activities of firms; the growth of cooperative ventures and alliances between, and within, the main wealth-creating institutions, the liberalization of both internal and cross-border markets; and the emergence of several new major economic players in the world economy' (Dunning, 2000: 8).

In parallel, a whole new ecology of knowledge has come to be recognized as part of the 'learning economy'. Following Lundvall (1992), this includes mastery over the know-what of facts and information, the know-why of principles that explain, the know-how of competence and skills, and the know-who of knowledge in networks of collaboration and communication. This spectrum calls into play a broad set of capabilities, from scientific knowledge and technical know-how and the education, skills and experience to apply and transmit them, to the social capital in inter-personal and inter-firm networks that provides the shared culture for learning in doing. These knowledge assets, and the various infrastructures that support and combine them, are now seen as the key inputs for innovation, learning and skill enhancement (rather than the other way round, as previously thought).

Knowledge creation has become the buzz-word for corporate success, and with this, accompanied by growing awareness of its character as an iterative process, a continuum of codified and tacit knowledge, an interactive and relational asset, and distributed within the firm as well as in links with others along the supply chain (OECD, 2000). This awareness, however, has also made it clear that creating and managing knowledge is a major corporate challenge, for knowledge often lies outside of the control of the firm, and is intangible, sticky, recursive and difficult to transmit or generalize.

For such reasons, a vast new literature has emerged on learning organiz-ations and organizational learning. It is not my intention to cover this literature. Instead, I wish to focus on one important theme at the heart of the new writing that tends to emphasize the anthropomorphic sources of knowing and learning. This concerns the separability of cerebral/cybernetic knowledge from practical/embodied knowledge, and whether particular states of the former can lead to superior forms of corporate innovation and learning. An influential line of reasoning in organizational theory that draws on cognitive psychology explains corporate knowledge in terms of the 'mental models' of individuals and collectives in organizations. This is evident in the work of Hayes and Allison (1998) who mobilize Argyris and Schon's (1978) pioneering distinction between single loop learning, defined as the changes in subjective theories or mental models within an existing paradigm (thus allowing better exploitation of existing know-how and expertise), and double loop learning, defined as reflection upon what has been learnt and deliberate questioning of core assumptions, leading to exploration beyond the paradigm. This distinction between single loop and double loop learning has become a commonplace in organizational research and is frequently invoked to explain the difference between corporations that make path-dependent innovations and those that generate novelty of a path-breaking nature.

Different rationalities are seen to be at work (Amin and Cohendet, 1999). Hayes and Allison (1998) explain both individual and collective learning repertoires in terms of cognition. Individual learning, thus, is related to variations in 'cognitive style', defined as 'a person's preferred way of gathering, process-ing, and evaluating information (p.848). This leads them to identify 22 different dimensions of cognitive style, where an organization 'stores the knowledge that is accumulated over time from the learning of its members in the form of an organizational code of received truth' (p.853). Thus, a common mind-set is established (through routines, rules, norms and habits of knowing and communicating), one that drives corporate learning.

But, is this how organizations learn? Can we narrow individual learning to particular mental dispositions and cognitive styles, and even if so, does collective learning work in this way? According to Popper and Lipshitz (1998), organizat-ional learning is not an extension of individual learning, in that organizations and their members lack the typical wherewithal for undertaking cognition on anthropomorphic terms. They suggest, instead, that learning by organizations is the result of purposeful structures and cultures that encourage learning, rather than cognitive dispositions at group or individual level. They also suggest 'that effective organizational learning is contingent on establishing a culture that promotes inquiry, openness, and trust', which they break down further into five 'hierarchically arranged values': continuous learning at the apex; the availability of full, undistorted and verifiable information; transparency or the 'willingness to hold oneself (and one's actions) open to discussion'; issue orientation rather than received wisdom; and accountability or 'holding oneself responsible for one's actions and their consequences' (p.170). While, clearly, the individual mechanisms identified can be debated, the thesis is clear: learning is an instituted process.

We can go much further, however, into the ontology of learning itself. There now exists a body of research on the fine-grained processes of knowledge creation in organizations that resonates with non-cognitive accounts of knowing. In developmental psychology, for instance, the pioneering work of Jean Piaget with children has shown that 'intelligence is internalized action and speech, and that both knowledge and meaning are context dependent' (Nooteboom, 2000: 2): we know and understand through practical action and speech acts, which, in turn are highly situated. Then, work in cognitive psychophysiology now weaves cognition into sensory and bodily functions, explaining that neural processes develop and are 'enacted' through the latter (Maturana and Varela, 1980; Lakoff and Johnson, 1999): we know through embodied neural repetitions why it is possible to speak of the powers of reflexive skills built into the tactility of the hand. Further, there is the suggestion from actor network theory that novelty is the product of connections, held in place by networks that combine an array of human and non-human elements:

> Innovation is by definition an emergent phenomenon based on gradually putting into place interactions that link agents, knowledges, and goods that were previously unconnected, and that are slowly put in a relationship of interdependence: the network, in its formal dimension, is a powerful tool for making these connections, and for describing the forms that they take. What marks innovation is the alchemy of combining heterogeneous ingredients: it is a process that crosses institutions, forging complex and unusual relations between different spheres of activity, and drawing, in turn, on inter-personal relations, the market, law, science and technology. (Callon, 1999: 2, translation of original in French)

This perspective on knowledge and innovation as the product of embodied practices finds ample support in research on the sociology and anthropology of corporate innovation. Work on both incremental and radical innovation in firms reveals that knowing and learning occurs through the daily practices of distributed communities of actors and non-human actants. As Brown and Duguid (1996: 76) suggest:

> Alternative worldviews ... do not lie in the laboratory or strategic planning office alone, condemning everyone else in the organization to a unitary culture. Alternatives are inevitably distributed throughout all the different communities that make up the organization. For it is the organization's communities, at all levels, who are in contact with the environment and involved in interpretative sense making, congruence finding, and adapting.

It is from any site of such interactions that new insights can be coproduced. If an organizational core overlooks or curtails the enacting in its midst by ignoring or disrupting its communities-of-practice, it threatens its survival in two ways. It will not only threaten to destroy the very working and learning practices by which it, knowingly or unknowingly, survives. It will also cut itself off from a major source of potential innovation that inevitably arises in the course of the working and learning.

COMMUNITIES AND LEARNING IN DOING

Learning occurs in communities, but there are different kinds of communities, which vary in remit, organization, and membership. For example, epistemic communities are groups of peers working explicitly on a common knowledge problem (e.g. scientists in a collaborative network). Knowledge production is an intended outcome and learning is based on mobilizing variety as the basis for experimentation, but within the frame of an accepted authority (e.g. a professional code) that binds the community and acts as a knowledge reference point. In contrast, communities of practice are homogeneous groups of employees engaged in the same practice in regular communication with others, often through mutual commitment (e.g. repair engineers or insurance claims processors). They are bound by common skills and tend to learn as a by-product of working and socializing together.

Firms can be seen as constellations of diverse communities of learning – sites where knowledges are formed, practiced and altered. These communities might be found in traditional work divisions and departments, but they also cut across functional divisions, spill over into after-work or project-based teams and straddle across networks of cross-corporate and professional ties. A common-ality, however, is that learning appears as an outcome of emergent meaning and novelty based on enrolment and engagement – as learning in doing.

The dynamic of unintended learning is exemplified in Wenger's (1998) brilliant longitudinal study of learning and innovation rooted in the daily practices of insurance claims processors in a large corporation. For Wenger, such communities of practice are marked by three dimensions of repeated interaction: mutual engagement, joint enterprise and a shared repertoire (of stories, artefacts, discourses, concepts against which existing and new practices can be evaluated). Wenger identifies three respective infrastructures of learning. The first infrastructure is engagement, based on mutuality, competence and continuity (e.g. locked in data, documents, and memory). The second infrastruc-ture is alignment, based on convergence (e.g. common focus, shared values), co-ordination (helped by standards, feedback, division of labour), and arbitration (facilitated by rules, policies, conflict resolution techniques). The third infrastruc-ture is imagination, composed of orientation (helped by examples, organiz-ational charts), reflection (supported by retreats, time-off), and exploration (facilitated by scenario building, prototypes, simulations, experimentation). These varied organizational orientations and tools that structure daily practices are said to provide the socio-technosphere that supports learning. The infrastructures, potentially, have enough novelty, perturbation and emergence in them to sustain incremental and discontinuous learning, as well as procedural adaptation and goal monitoring.

A similar sociology of learning can be found among dispersed communities of practice. An example is Orr's (1996) celebrated study of Xerox technical reps who service and repair the corporation's copiers at customer sites. Orr discovered that the information and training provided to the reps, who tend to work alone, was inadequate for all but routine tasks. Instead, the forcing ground for new learning was avidly frequented occasions such as breakfast or lunch, when reps would come together, to mingle eating and idle chat with endless talk about work – discussions of problems, solutions, experiences, customers and technical standards. Similarly, intractable problems were often resolved after hours of talk during which stories of past problems were exchanged and gradually respective repertoires were aligned. Significantly, such learning, 'off the road and without maps', is facilitated by sociality, drawing on the powers of collaboration, narration and improvisation: 'talk and the work, the commun-ication and the practice are inseparable' (Brown and Duguid, 2000: 125).

The processes outlined by Wenger (1998) and Orr (1996) also apply to strategists, scientists and 'big' discoveries. An early insight of studies in the sociology of science was that knowledge practices in R&D laboratories were not that different from those underpinning incremental innovation in, say, the Breton fishing industry. In the laboratory too, the production of novelty drew upon routines, conversations, meetings, scripts, memory, stories and other soft technologies, grafted onto the technologies of formal knowledge acquisition and application (Latour and Woolgar, 1979; Latour, 1986). The inseparability of the knowledge chain, and recursivity with it, makes it virtually impossible to assign the moment of discovery to a sudden brainwave, a retraining programme, a conversation at a coffee break, or a technology-driven prompt.

This is also often the case with accidental discoveries. A revealing example is Hutchins' (1996) study of how a navigation team arrived at a new procedure when, upon entering a harbour, a large ship suffered an engineering breakdown that disabled a vital piece of computer-based navigational equipment. Following a chaotic and unsuccessful search for a solution through thought experiments and computational and textual alternatives, the team developed an answer through doing. As local tasks were found for individuals distributed across the ship, the ensuing sequence of actions and conversations, drawing on experience and experimentation, led to the construction of a solution based on trial and testing. Although Hutchins describes what happened as 'cognition in the wild', the point is that it involved engagement and enactment.

Where, then, might we locate the difference between incremental and radical learning, or between the exploration and exploitation of novelty? For Callon (1999), it lies in the nature of actant networks, that is, in the terms on which engagement occurs. He distinguishes, for example, between stable and emergent networks. In the former, innovation objectives and goals are known, and the emphasis is placed on combining known complementary competences with codified or tacit knowledge in pursuit of programmable action. Network stability, in terms of composition, duration, replication and length, is a key property. Emergent networks, in contrast, yield identities, interests and competences as the outcome of 'temporary and experimental translations [*traductions*]' (Callon, 1999: 29). In emergent networks, actors with different knowledges come together into an uncertain venture, with ill-specified objectives (experimentalism is the goal). In order to work, equivalence (linguistic and cultural) needs to be established between interests that are distant, incommensurate, unstable and uncertain. New knowledge is the product of interaction, learning and adaptation as you go along, involving trial and error, gropings, but the practices of translation and enrolment play a crucial role in supporting the production of experimental knowledge.

ORGANIZING TO LEARN

The idea of learning in doing raises interesting questions about the organization and management of corporate innovation and creativity. The knowledge-based approach to the theory of the firm has tended to focus on coordination mechanisms and issues of organizational design (e.g. rules and directives, time-sequencing and routines as 'grammars of action'). The ethnographic approach takes us in a different management direction, into the realm of how the practices of engagement/enrolment/translation can be supported. This difference springs from its interest in the process itself of how knowledge is formed and made ex-plicit, something that the knowledge-based perspective tends to take for granted.

A starting point is the recognition itself that organization is a means of enacting the environment, and through this, learning through simulation. Vicari and Toniolo (1998) have argued that firms make sense of the environment only from their enactments of it, with the help of varied 'cognitive schemata' (e.g. scripts, maps, data and stories). They only have knowledge of themselves, but equally, they need to produce new knowledge in order to survive. According to Vicari and Toniolo, this is attempted via the construction of images of reality based on interpretations of raw data, expectations of customer and competitor behaviour, time and motion studies, and so on. The performance of these enactments is the basis upon which firms know and act. One implication of the proposition that market search occurs through enactment is the need for explicit recognition of, and organization for, learning through the 'production of errors' (Vicari and Toniolo, 1998: 211).

But, we can be more explicit about organizing for learning in doing. For Brown and Duguid (1996), an imperative is the encouragement of a varied ecology of communities of practice across an organization because 'out of [the] friction of competing ideas can come the sort of improvisational sparks necessary for igniting organizational innovation' (p. 77). The argument is that a varied selection environment for innovation (with which most evolutionists would agree) allows both learning along different pathways and the possibility of new learning through friction. Management by design of learning (e.g. via training courses, acquisition of R&D, new technologies and new competences) is not sufficient for the production of novelty. Communities of practice, though fundamentally informal and self-organizing, need to be recognized, supported and valued.

Learning in doing requires some underpinning of the 'soft' infrastructure for learning. For example, supporting slack, memory and forgetting is of crucial importance. Slack, namely the retention of skills and capabilities in excess of those needed for immediate use, helps 'innovative projects to be pursued because it buffers organizations from the uncertain success of these projects, fostering a culture of experimentation' (Nohria and Ghoshal, 1997: 52). Memory, facilitated by remembering, inter-generation mixture and stories among employees, mobilizes the fruits of experience, including knowledge of past trials and errors. Forgetting, through employee rotation, new training and new routines, helps to weed out practices not suited for changed circumstances. Another element of the soft infrastructure, so clearly evident from the work of Wenger (1998), Orr (1996), and Brown and Duguid (2000), is the practice itself of community, through support for the daily practices of a group held together by shared purpose and expertise (e.g. insurance claims processors, middle managers and R&D workers). This can be done through a range of technologies engagement, including 'learning in talk' at away-days, regularly slotted meetings and (tele)conferences, devices and spaces that facilitate translation and enrolment between dissonant knowledges, out-of-hours events and recreation opportunities to underpin sociality: the engineering of talk.

A third and crucial element is attention to dissonance and experimentation. Creative communities are those that are able to mobilize and confront difference and disagreement. Learning within them is a matter of exploiting existing competences, but it is also a matter of retaining variety so that new opportunities are not lost, and of managing dissonance as a source of experimental knowledge. Firms need to engineer exploration and standard practices here include using competitions between work groups for new ideas, intermediaries to help divisions adopt new routines, scenario building and experimental games. But there is more. According to Thrift (2000; also in this volume), an entirely new management culture valuing creativity in its own right is arising, with firms increasingly drawing on cultural projects and performativity as a means of encouraging innovative thinking and practice. These include the re-engineering, in new business magazines, of business professionals as fast, youthful and urbane figures, the growth of management training courses that get participants to perform in plays, stories, historical enactments and musicals to unlock their imagination, and the deployment of drama and speech techniques by business consultants to broker confidence, intent and mutuality among clients.

Finally, in a system of distributed learning, a key management challenge is to hold the network in place and to align autonomous centres of innovation towards core business goals and priorities. The rules of hierarchical manage-ment, based on task specialization and the division of responsibilities along the chain of command, are largely inappropriate because local units are self-organized and composite in nature. There is a large volume of literature on the management of heterarchical and decentred firms and their complex network arrangements. This is not the place to review the literature. Instead, I want to conclude this section by focusing on one management imperative that springs out of an anthropological reading of knowledge formation. This concerns the nature of what might hold a network of distributed learning in place.

The earlier emphasis on knowing through talk implies the centrality of securing effective communication between various knowledge/language commun-ities because in a decentred organization, 'the real leverage lies in creating a shared context and common purpose and in enhancing the communication densities within and across the organization's internal and external boundaries' (Nohria and Ghoshal, 1997: 87). Interestingly, in choosing the appropriate management devices, Nohria and Ghoshal emphasize the role of socialization (e.g. via corporate encounters, conferences and recreational clubs), normative integration (e.g. via incentives such as access to healthcare or travel concessions, company rituals, inculcation of corporate or brand standards) and effective communication between self-governing units (via both internet and relational or cognitive proximity).

A similar account is given by other influential business analysts. Ichigo et al. (1998) argue that tacit knowledge can be made less sticky through the use of 'knowledge enablers' (e.g. on-going dialogue with customers, personnel exchanges with suppliers). Probst et al. (1998) highlight the role of 'languaging' devices that enable knowledge integration across boundaries (e.g. through in-formal networks in associations and clubs, employee exchanges between firms). Brown and Duguid (1998: 104) note the role of 'in-firm knowledge brokers', working with overlapping communities in order to loosen strong internal ties that restrict exploration, and 'boundary objects' such as contracts, plans, blueprints and other technologies and techniques that help 'make a community's own presuppositions apparent to itself, encouraging reflection and 'second-loop' learning'.

SPACES OF KNOWING

Brown and Duguid (2000) see the local clustering of firms possessing complementary and rival knowledges as an important organizational means for corporate learning. For them, such local clusters are 'ecologies of knowledge', replete with varied and redundant knowledge, with spatial proximity keeping firms on their toes as a consequence of the circulation of ideas in local associational networks, local labour mobility, tracking of rivals and the high standards resulting from local product specialization.

In fact, for some years now, economic geographers and evolutionary economists have claimed that tacit knowledge is locationally sticky and that firms are embedded in national and regional innovation systems, against the exaggerations of economists insensitive to geography as a habit of mind (e.g. via claims about context-free economic actors) or by empirical persuasion (e.g. via claims about the effects of globe-spanning technologies). Their argument centres, on the one hand, around the claim that tacit knowledge is formed relationally, and for that, is 'context-dependent, spatially sticky and socially accessible only through direct physical interaction' (Morgan, 2001: 15). It also draws on the claim that all firms, including TNCs, draw on and are shaped by the institutions of the 'home base' at national or regional level as a resource for technological and non-technological innovation through the R&D system, the education and training system, financial and academic links with industry, business norms and practices, associational arrangements and non-market interdependencies (see Dicken, Gertler and Malmberg, in this volume).

Trenchantly, Morgan (2001: 14) has rebuked those who 'devalue the significance of geography', including commentators like myself who ask if relational proximity can be reduced to spatial proximity (Amin and Cohendet, 1999; 2001; Oinas, 2000). The key question, though, is what should we take geography to mean in this context? Do we have to follow a geography of points, lines, and boundaries, reduced to the opposition between, on the one hand, *place* as the realm of near, intimate and bounded relations, and on the other hand, *space* as the realm of far, impersonal and unbounded relations, such that the tacit/contextual nature of learning can only be recognized in terms of spatial/physical proximity? Is it not possible to imagine another – topological – geography, made up of organizational networks of varying length and spatial composition, network sites of varying intensity of proximate and distant connectivity, proximity that is also of a non-territorial nature, and mobilities and flows that count as more than spaces of transit? It is the tracings of this alternative geography that I wish to develop in the rest of this section, with the aim of claiming that even the most relationally proximate form of learning, namely learning in doing within communities, involves more than 'being there' defined as spatial proximity – face-to-face contact, local ties, 'the 'home' base and the like.

Crucially, if it can be shown that the sociology of learning is not reducible to territorial ties, there is no compelling reason to assume that 'community' implies local community, or that local ties are stronger than interaction at a distance. Of course, many communal bonds *are* localized, as we might find in a neighbourhood-watch scheme or a community of practice consisting of employees in a given site, but many other communal bonds – of no less commitment and intensity – draw on a spatially 'stretched' connectivity. This includes communities of enthusiasts with like interests (e.g. vegetarians, DIY groups, road protestors, clinical psychologists) held together by cheap travel, the internet and specialist literature. It includes tightly knit diasporas communities that are based on kinship and migratory ties around the world as well as shared imaginary cultures that do not correspond to particular locations. It includes the community of sympathy for others around the world, mobilized by the media through reports on innocent war casualties.

To return to the theme of corporate learning, I believe that Nonaka and Konno's (1998: 40) use of the Japanese philosophical concept '*ba*' (roughly translated as 'place' in English), indicating a 'shared space for emerging relationships' as 'a foundation in knowledge creation' can help us acknowledge the plural geographies of interaction and connection. For Nonaka and Konno, *ba* 'can be physical (e.g. office, dispersed business space), virtual (e.g. e-mail, tele-conference), mental (e.g. shared experiences, ideas, ideals), or any combination of them' (p. 40), which, in turn, sustain four kinds of knowledge space. First, '*originating ba* is the world where individuals share feelings, emotions, experiences, and mental models ... Physical, face-to-face experiences are the key to conversion and transfer of tacit knowledge' (p. 46, original italics). Second, '*interacting ba* is more consciously constructed ... Selecting people with the right mix of specific knowledge and capabilities for a project team, taskforce, or cross-functional team is critical. Interacting *ba* is the place where tacit knowledge is made explicit ... Dialogue is key for such conversions' (p. 47, original italics). Third, there is '*cyber ba* ... a place of interaction in a virtual world... Here, the combining of new explicit knowledge with existing inform-ation and knowledge generates and systematizes explicit knowledge throughout the organization' (p. 47, original italics). Finally, '*exercising ba*'... facilitates the conversion of explicit tacit knowledge to tacit knowledge ... enhanced by the user of formal knowledge (explicit) in real life or simulated applications' (p. 47, original italics).

Knowledge spaces, more accurately, proximities of knowledge, are not reducible to matters of geographical distance. There is clearly an overlap be-tween the two, but knowledge spaces cannot be read off geographical properties as has been the tendency in recent economic geography (e.g. the special powers of local face-to-face contact, the loss of intimacy that come with distance, and the largely formal or codified nature of distant knowledge links). It could be argued, following this insight, that part of the history of modern organization has been to facilitate proximity at a distance, so that corporations can effectively mobilize knowledge in different places, transmit it over a distance, and translate, transform or embed it in its travels through the corporate network. One of the purposes of corporate form, or the rules and practices of technological ordering and spatial distribution, and the conventions of communication, command and control, is to hold varied knowledge architectures in place and to establish relational proximity and knowledge coherence across different spatial scales. This focus on organization as a means of maintaining and managing a differentiated spatial ecology of knowledge, through complex network formation and network management devices, allows recognition of a more varied geography of knowledge and learning including 'being there' in distributed communities, the travels of know-how generated by central management and R&D labs, knowledges carried by travelling executives and technicians, embodied in varied technologies, transmitted across space, and the insights generated through occasional meetings, teleconferences and telephone conversations or email messages sent in transit. Through the varied architectures of modern corporate organization, a rich spatial ecology of knowledge has become rendered as domestic or relational knowledge and 'being there' is no longer a constraint of geographical proximity or a special property of place.

Let us look at some examples of the varied uses of space for learning by large corporations aided by the muscle of distributed organization. For example, Schoenberger (1999: 216) notes that when the firm 'realizes it needs to change', it can 'consciously set out to create a new kind of place within the firm', including 'organizational and geographical separation from the centre'. It can engineer a dislocation that is at once centred and decentred. Hatch (1999) makes a similar point concerning 'empty spaces' in organizations to mark a space for creative action that is not regulated by rules and norms: a kind of working and thinking space that is not reducible to a location – here or there. In McKinsey and Andersen Consulting, for example, project development teams 'are not traditionally co-located or isolated from the rest of the firm', but 'consist of consultants working out of offices across the country' because 'it gives the team access to the most resources within the organization because each individual team-member knows and regularly talks to their colleagues back at the office' (Hargadon, 1998: 218).

The burgeoning literature on spaces of creativity illustrates the varied geographical 'forcings' (in the sense they are both organized and dynamic) at work. First, as Grabher (2001) suggests from his work on the Soho advertising cluster in London, creativity is the creature of heterarchic organization – of diversity, rivalry and incoherence negotiated, always only temporarily, through communities of practice and projects with shared goals. The geography of 'heterarchic novelty' includes face-to-face meetings, sociality and casual contact in the knowledge-rich 'village' of Soho, but it also draws on distant objects such as drawings faxed between offices around the world, global travel to form temporary project teams, and daily internet/telephone/video conversations with distant clients and collaborators.

Second, therefore, mobility is built into the knowledge system. As Thrift (2000: 24) notes, 'new means of producing creativity and innovation are bound up with new geographies of circulation which are intended to produce situations in which creativity and innovation, can, quite literally, take place'. Thrift mentions three of these geographies of circulation: the 'constant quartering of the globe by executive travellers' (p. 24) to be at meetings where face-to-face contact permits brainstorming and inspiring passion; the 'construction of office spaces which can promote creativity through carefully designed patterns of circulation' (p. 26), such as the replacement of offices by 'hot desks' that busy employees can book for a finite period, or the creation of common spaces to ensure serendipitous contact; and the engineering of 'virtual clustering' via such experiments as internet 'thinking studios' or 'innovation exchanges' which 'promote knowledge exchange and sharing by bringing different actors together' (p. 28). Circulation has become a deliberate means of maintaining the 'potential for creating unexpected connections' (Hargadon, 1998: 219).

Third, all manner of actants – situated, enrolled and in movement – make up a network topology that, thus defined and in its entirety, supports creativity. This is the great geographical insight of actor-network theory, replacing as it has, the geography of pre-defined places and boundaries, with a geography of sites, containments and contours that unfolds through purposeful acts. So, it is perilous to claim that the creative phase of an innovation cycle has to be localized because of the need for face-to-face interaction during experimentation (Sölvell and Bresman, 1997), for this is to ignore the multiple object geographies (of scripts, scribbles, scientific conventions, rail journeys, internet exchanges and world conferences) that facilitate the enrolment of different actants in emergent networks (Callon, 1999).

Modern science certainly works in this way. 'Molecular biology', notes Rabinow (1996: 24), 'has taken up the current conjuncture through an increased use of electronic means of communication, of data storage, of internationally coordinated projects like the human (and other organisms) genome projects. The circulation and coordination of knowledge has never been more rapid or more international.'

Maybe it has always been like this. In an insightful paper on scientific discovery in sixteenth-century Europe, Harris (1998) shows that 'long-distance corporations' such as the Spanish colonial House of Trade, the Dutch East India Corporation, and the Society of Jesus – all possessing considerable global reach – played a critical role in instituting medical, cartographic, biological and botanical discoveries through a 'kinematics of scientific practice' based on the to-and-fro of knowledgeable people and a vast variety of knowledge objects such as maps, quadrants, dials, chronometers, compasses, logs, descriptions and correspondence. As Harris (1998: 271, original italics) explains:

> ... the long-distance corporations – that is legally constituted corp-orations that had more or less mastered the operation of long-distance networks – had immediate institutional need of certain forms of natural knowledge and therefore incorporated knowledge-gathering and know-ledge-producing mechanisms into their social fabric. These practices, though directly related to academic and bookish disciplines we think of as scientific, were situated in the *vita activa* of a corporation's member-ship and were necessary tools in the prosecution of corporate agendas. Moreover, the dedicated channels of communication required in the operation of long-distance corporations facilitated the movement of personnel, texts, and objects in both the pre- and postproduction phases of knowledge-making ... the long-distance corporation is an especially promising research site for the integration of local or embedded knowledge on the one hand and geographically distributed practices on the other.

The modern transnational corporation is typically one of these long-distance corporations, now doing most of the above within its own corporate boundaries and other networks under its tutelage. It perhaps has a greater knowledge base under its control and for that it may well face distinctive management challenges, such as how to bend path-dependencies and inertia based on accumulated knowledges, how to manage a system of distributed knowledge, how to reconcile knowledge and learning imperatives with routine transactional imperatives and how to balance exploration of new knowledge with exploitation of existing knowledge. But, like the long-distance corporation of old, it draws upon a fine network architecture of connections and mobilities and a whole array of governance technologies to make sense of varied and often conflicting knowledge domains, each of different spatio-temporal reach. Like the old long-distance corporation, it achieves relational proximity through trans-lation, travel, shared routines, talk, common passions, base standards, brokers, epistemic and community bonding, and the ordering and orientation provided by files, documents, codes, common software and so on.

In summary, there are many geographies implicated in the production of novelty, many ways of establishing relational proximity. As Harris (1998: 296) notes, there is no compelling reason to accept that when 'one speaks of 'local', 'situated' or 'embedded' knowledge, the implication' must be 'that the narrative is somehow confined to a small 'space' – if not in the literal sense of a geographic metric, then at least in the sense of a restricted social, cultural, and temporal metrics'. Corporate organization, with all its tools and brokering devices, provides the mundane means behind a mobile knowledge space made up of bits and pieces from all over.

CONCLUSION

The literature on the everyday sociology of knowledge that I have traced in this chapter has highlighted two controversial claims on corporate learning. The first is that a substantial amount of learning is of a non-cognitive nature, situated in the practices of communities; practices, in turn, that are the source of both radical and incremental innovation. This emphasis on situated practice helps to displace the current hyperbole on the centrality of knowledge workers, big science, information highways and reflexivity in the knowledge economy of the future. An understanding of the inventiveness of communities of practice – be they insurance claims processors or engineers and scientists – shifts the debate in a different direction by appreciating the centrality of 'boring' old things such as socialization, sociability, work and the practices of doing for knowledge creation. In the new hyperbole, work practices seem to have become reduced, at best, to transactional necessity, and, at worst, to a troublesome source of class solidarity, unionism and insurgence. Such a reading must be spurned, for it is wrong to assume that, like Superman, the free-floating knowledge worker alone has the remarkable powers to renew capitalism.

The second controversial claim that emerges from the sociological literature is that the space of intimacy and familiarity – seen increasingly to be vital for learning – is varied in geographical composition and certainly not reducible to being in one place. Thus, to recognize that:

> ... the global economy is constituted by 'spaces of network relations'... embedded in particular spaces ... does not mean at all that all social actors in each network must be bound together in exactly the same territory. Rather ... there are 'spaces' for social actors to engage in network relationships. These 'spaces' can include localized spaces (for example financial districts in global cities) and inter-urban spaces (for example webs of financial institutions and the business media that bind together global cities). The global economy is thus made up of social actors engaged in relational networks within a variety of 'spaces'. (Dicken et al., 2001: 97)

Such an analytic opens up a new way of doing the geography of corporate learning. We might, as Dicken et al. (2001) suggest, look at what goes on in localized and inter-urban spaces. But we can do more. We could map the varied geography of network tracings that make up a community of knowledge and measure the effects of objects at a distance on what goes on in any one site, or the effects of local conversations on standardized knowledge. We could do a geography of mobilities and placements in order to get a better sense of how far knowing is dependent on travel, virtual communication, special meetings, short-hops, away-days, knowledge brokers, consultants and drama workshops. We could, finally, embark on a geography of recursivity to see what role spatial iterations play in the to-and-fro between trial and error, serendipity and deliberation, tacit and codified knowledge that seems so crucial for creativity.

ACKNOWLEDGEMENT

This chapter draws on collaborative work with Patrick Cohendet (Amin and Cohendet, forthcoming). I wish to thank Henry Yeung and Jamie Peck for their comments on the first draft and also my doctoral student Jong-Ho Lee for introducing me to literature on organizational learning that I was not familiar with. Finally, a debt to Peter Dicken for showing the value of a light touch when dealing with hard corporations.

Chapter 8

THE MIGHT OF 'MIGHT':
HOW SOCIAL POWER IS BEING RE-FIGURED

Nigel Thrift

INTRODUCTION

In this chapter, I want to consider how the exercise of capitalist power is changing in contemporary Euro-American societies through the agency ofa large-scale reformatting of space and time. I want to produce a picture of a particular kind of capitalism, one which is both more and less controlling, and a capitalism that takes the world as it makes it. In attempting to paint on such a broad canvas, I run the obvious risk of producing generalizations that hold everywhere and nowhere. But I hope that by pointing to the central role of new forms of location in this reformatting (though what the exact meaning and use of 'location' might be is clearly open to question) I can avoid at least some of the pitfalls of such an abstract approach.

Most particularly, I want to concentrate on the way in which locations are increasingly being *designed* to proliferate, but to proliferate *adaptively* (rather than as mere copies). This process of adaptive radiation in turn allows new and inventive configurations of 'might' to surface; new ecologies of possibility, if you like. But this proliferation of possibility[1] is not a random or uncontrolled process. It involves the co-evolution of new forms of creativity *and* new forms of standardization which are able to both boost the process of inventiveness and also control it, through continuous monitoring and adaptation; the listening and reacting to feedback that Deleuze (1993) calls *modulation*.

The proliferation of these new practices of capitalist power (the might of 'might', as I call it) is not accidental. It arises from the concatenation of a series of different processes which can be interpreted as the result of the adoption by capitalism of a new kind of transcendental empiricism and the consequent casting off of classical notions of empiricism (which argue from a logic of abstr-action and generalization) for a very different 'synthesis of the sensible', one based upon the idea of economy and society as a continuous experiment. Or, to put it another way, this is a move from considering states of things as unities or totalities to considering them as multiplicities, 'a set of lines or dimensions which are irreducible to one another' (Deleuze and Parnet, 1987: 111).

I am not suggesting that this shift to a new kind of empiricism is a conscious plot. Rather, it derives from the practical logics inherent in three different but related processes which in large part both arise from the agency of capitalist firms and also provide the momentum for these firms to keep changing. One process is *discursive*, and consists of the increasing use of biological analogy, founded in notions of evolution. The use of such analogy has now returned in full force, its power boosted by the ubiquity of genetics. It can be seen at work theoretically in the renewed popularity of Bergson's vitalism, as seized upon and reworked by Deleuze. And it can be seen at work practically in attempts as diverse as the construction of biological models of business (Clippinger, 1999), and the construction of new, friendlier 'information ecologies' (Nardi and O'Day, 2001) that can light the way to new kinds of socio-spatial. Whatever the attempt, the biological metaphor is used to suggest a creatively adaptive evolutionary tendency borne out of new combinations of hybrid materials that can 'self-design' (Weick, 2001).

The second process is *organizational* and consists of the derivation of a new heterarchical logic of organization which is not based upon a 'one size fits all' stabilized and reproducible order but upon open architectures which promote diversity. Organizations must invest in forms that allow for easy reconfiguration by balancing adaptive fit to the current environment with the need for adaptability in subsequent dislocations through the promotion of coexisting logics and forms of action. They are then able to foster long-term adaptability through better 'search' – 'better because the complexity that it promotes and the lack of simple coherence that it tolerates increase the diversity of options' (Girard and Stark, 2002: 3).

The third process is *technological*, and consists of the rise of a new informational ethology, based upon the pervasiveness of inter-communicating software. This ethology is beginning to produce an archive of information about everyday life, and, as importantly, a set of skills of manipulation of this archive, that are sufficiently persuasive and persistent that they can act as the foundation of a new, continuously modulated grid of everyday checks and balances. In turn, this archive has begun to affect behaviour in the most effective way possible – by acting as a taken-for-granted background.

This chapter pursues this new capitalist empiricism – and its consequences – by attending to the ways in which it is producing new forms of location, *new spacings and timings*. Whilst many of these spacings and timings are still hesitant projects, I think that, when placed together, they begin to make it possible to see just how pervasive practices of modulation – with all their attendant (and fruitful) overcodings – are now becoming.

The chapter is therefore in four parts. In the first part, I will consider a set of mobilizations of space and time currently being driven by capitalist firms which I take to be symptomatic of the new modulated empiricism. Next, in order to head off too delirious an interpretation of the new empiricism, I will therefore add in three new coeval processes of standardization that are currently taking place – new forms of writing, the increasing relevance of the affective, and the recoding of built forms – which cross-cut these spacings and timings in ways which show that control is increasingly modulated – but is still control. In the penultimate part of the chapter, I will turn to the response of capitalist firms to these new conditions of space and time that they have in large part created. Then, I will offer some brief concluding comments, in which I will suggest that what we can see is that the new context-specific forms of standardization that have started to become possible have begun to produce judgements of worth (Thevenot, 2001) that would have been very difficult to enact before.

NEW SPACINGS AND TIMINGS

How is it possible to understand contemporary capitalist spacings and timings? In what follows, I will point to four symptomatic developments, each of which betrays the ambition of capitalist firms to engineer space and time at finer and finer grains in order to produce larger and larger gains. The first of these developments is symbolized by the recent attempts to produce see-through concrete, with the hope of producing 'cities that glow from within, and buildings whose windows need not be flat rectangular panes, but can be arbitrary regions of transparency within flowing, curving walls' (*The Economist*, 2001: 2). The search after such a material is symptomatic, I think, of a need to produce operational spaces in which the work of interrelation is completely visible. It is the product of a sense that everything can and even must be seen. Much of the push for this kind of transparency probably arises from the increasing pervasiveness of mediated entertainment and communication, resulting from their thorough embeddedness in all aspects of everyday life, so providing a dense network of screened and otherwise transmitted elsewheres which are both site-specific and echo off each other (cf. de Nora, 2000). In such spaces and times, the panoptic gaze of previous times is being replaced by something much less focused which depends on the very ubiquity and mundanity of mediated seeing. The clamour for attention produces a flickering, trawling gaze in which attention 'is seized and dropped and held and released by possibilities of meaning that amuse and interest but do not quite come into being' (Bromell, 2000: 133).

Of course, none of this is to say that nothing is systematically surveilled or hidden. Rather contemporary timespaces are made up of greater areas of free play which are, *at one and the same time*, constantly monitored, allowing events to be zeroed into once they reach certain attention levels. Similarly, modern instruments for knowing the world consist of large areas of free play, but still contain kernels unable to be accessed, as in the locking up of many privileged input and output commands in software (Kittler, 1997; Lessig, 2002), or the new generation of closely specified genetic patents (Whatmore, 2002).

The second symptomatic development is the much greater attention being paid to the 'moment'. The ability to register, understand and engineer smaller spaces and times measured in microns and milliseconds is a critical element of modern cities at a series of levels, since it is one of the key means by which the world is currently being apprehended and made more of. Four different means of registering fleeting spaces and times have been invented.

First, there is the ability to sense the small spaces of the body (Amato, 2000; Stafford, 1996; 1999). Through the texts and instruments of science we can now think of the body as a set of micro-geographies. For example, 'in the past, the microscope exposed the thickness of experience, the depth of the level' (Stafford, 1991: 202). Today, the sub-visible body is tracked in other ways too. For example, modern computerized tomography has, through magnetic resonance imaging, been able to map the emergent landscape of neuronal firing (Le Doux, 1998, Damasio, 1999).

Second, the ability to sense bodily movement has become greater. Beginning with the work of Etienne-Jules Marey and moving into the age of media overload, we can now think of space as minutely segmented frames of time, able to be speeded up, slowed down, even frozen for a while (Dagognet, 1992). Third, numerous body practices have come into existence which rely on and manage such knowledge of small times and spaces, most especially those connected with the performing arts, including the 'under-performing' of film acting, much modern dance, the insistent cross-hatched tempo of much modern music and so on (Thrift, 2002). And special performance notations, like Labanotation, allow minute movement to be recorded, analysed, and recom-posed (as in much animated film). Then, finally, a series of discourses concerning the slightest gesture and utterance of the body have been developed, from the elaborate turn-taking of conversational analysis to the intimate spaces of proxemics, from the analysis of gesture to the mapping of 'body language', which, suitably packaged up, have made their way out in to the world (e.g. McNeil, 1992).

In other words, what we can see is what was formerly invisible or impercep-tible becoming constituted as visible and perceptible through a new structure of attention which is more and more likely to pay lip-service to those actions which go on in small spaces and times, actions which involve qualities like anticipation, improvisations and intuition. In other words, the second-to-second artfulness of the body (Katz, 1999) is now described and reconstituted as a *resource*.

We might say in summary that our structure of attention now involves the inhabitation of much smaller spaces and times than before, spaces and times which proceed out of the general assumption that perception can no longer 'be thought of in terms of immediacy, presence, punctuality' (Crary, 1999: 4). Per-ception is both stretched and intensified, widened and condensed.

We can see this new framing of perception in another way, of course; as the practical culmination of a whole strain of work on the psychology and physics of *attention* dating from van Helmholtz and Wundt's work in the nineteenth century, through the work of Kulpe, Tischer and Jones in the latter years of the nineteenth century, through the work of Libet and others in the mid-twentieth century, and so on (Pashler, 1998). It is no coincidence that what we see now is the large number of business books appearing which argue that 'every organization is an engine fuelled by attention' (Davenport and Beck, 2001: 17). And, in turn, this focus on attention has produced new forms of industrial technology: the brand, focusing on phatic communication, the experience economy, focusing on the production of entertaining experiences through minutely scripted events, and new conducts of business which involve dynamizing the whole business body (Thrift, 2002).

The third symptomatic development is the redefinition of spatial and temporal agency away from the body. Increasingly, the world is being run by transhuman actants in which flesh is often only a subsidiary element of action (Lash, 2002). Yet, paradoxically, that dependency has produced a greater and greater attention to the body. As the subject increasingly becomes just another object (Boyne, 2001), so the engineering of the body becomes ever more complete. Of course, the engineering of the body is hardly a new phenomenon, from army drill through Taylorist work principles, the body has been a key element of agency. But modern capitalist organizations are likely to work over a greater range of body movement, and are much more likely to pay attention to non-representational aspects of the body like affect which can only be shown up and worked with through the careful articulation of space and time. It is no coincidence that we live in the age of the so-called 'expressive organization', an organization which can work on and boost the allegiances and energies of bodies by conjuring up precisely the distinctive spatial-temporal formats in which these allegiances and energies can be disclosed (Thrift, 2000).

The fourth symptomatic development is a change in the conduct of space and time in everyday life created by the rise of mobile communications. In the growing literature on new forms of portable telecommunications like email and text messaging, this is called the growth of *hyper or micro-coordination,* whereby it is no longer necessary to take an agreement to meet at a specific time and place as immutable, and of *perpetual contact,* whereby it is possible to be in continuous contact with others: access becomes a way of life (Rifkin, 2000). Boosted by attendant technologies like GIS and GPS, email and text messaging have endowed societies with the ability to coordinate and re-coordinate at a dist-ance on an all-but-continuous and continuously adjusted basis. As a result of the advent of this increasingly persistent electronic timespace, the nature of social interaction itself is able to change. The new telecommunications are being used to continually modulate where and when encounters will take place in a kind of intricate ballet of circumstance of the kind that used to have to be reserved for public meeting places like the street (Brown et al., 2002; Katz and Aakhus, 2002).

In turn, these four developments are producing the beginnings of a new vocabulary of space and time. It is no surprise, as I have argued elsewhere (cf. May and Thrift, 2001), that there is a resurgence in vitalist ideas, an interest in generalizing notions of ecology and ethology, a wholesale plundering of nonequilibrium forms of chemistry and physics, or a turn back to various pragmatisms. What is being searched for is a means of describing timespace that cannot be encompassed by simplistic phrases like 'time-space compression' which assume a set of valences that can be fitted into a single uni-verse. Rather, the description is invested in a complex set of *skills* of adaptation to multi-verses which are both time and space-specific and count on time and space as the medium for their expression. Hence, the large degree of inspiration now being found by so many in site-specific art, in various new forms of plastic architecture, in different kinds of performance, and so on – for these are not simply aesthetic archives but also indicators of the kinds of adaptive-cum-creative skills that have now become more and more important to survive and prosper in capitalist societies.

NEW FORMS OF NORMALIZATION

These symptomatic developments may all sound remarkably voluntaristic. But I have tried to show that in each case inspiration is being measured out in quite deliberate ways. And, as I want to show in this section, these developments are being assisted by the growth of new complexes of normalization of time and space which because they are increasingly context-sensitive, are increasingly attuned to (and productive of) streams of locations within which the modes of experience, conceptualization or representation that are allowed can vary and yet still be controlled. Thus, much greater stress is now being placed on the continuous work of standardization or classification as the 'artful integration of local constructs, received standardized applications, and the re-representation of information' (Bowker and Star, 1999: 292), often incorporating any resistances and interferences by managing entities at more detailed (and interconnected) levels and by offering certain kinds of participation in the work of classification and standardization (as in open source software). At the same time, of course, standards and classifications are still often contested in these new complexes: low level conflicts rumble on concerning which classifications and standardizations are appropriate, as in the current conflict over location-specific standards for wireless. Below, I want to consider three of these new complexes of normalization in turn.

I want to start by considering new forms of *mechanical writing*. Writing, of course, has a long and involved history (see, for example, Fisher, 2001; Hobart and Schiffman, 1998). But mechanical writing is generally of a more recent vintage. Born out of the symbiosis of writing and the turn to information metaphors of the world found mainly in information theory and cybernetics, it now forms an unspoken backdrop to so much of the wider world. Just like the history of writing, it can be considered to have started with the use of token or graphic symbols that mark possession and, in particular, the bar code. Based on Morse code, the bar code was invented by Joseph Woodward and Bernard Silver in 1949, patented in 1952 and first used commercially in 1974. The bar code has subsequently proved to be one of the crucial inventions of the modern urban economy. Administered by the Uniform Code Council, which issues universal product codes, it is estimated that Universal Product Codes (UPC) are used by a million companies world-wide and these codes are scanned five billion times a day. But UPC codes constitute only about half of bar code usage. For example, large agencies like FedEx, UPS, and the US Postal Service use their own proprietary bar codes to move mail and parcels. Bar codes are the chief means by which inventory identification and control, the key to modern logistics – a subject to which I return below – is realized.

In the computer world, the equivalent of the bar code are network address locations like the.sig file. First invented circa 1980, probably on an online bulletin board like FidoNet, the.sig file provides a co-ordinate description of the sender job title, phone number and so on and a kind of digital soundbite fixed below this description. In turn, the sig file produced one of the most successful business strategies – Hotmail – web-based email which attracted more than 12 million users in its first 18 months. Now owned by Microsoft, Hotmail has more than 60 million subscribers.

But, clearly, these symbols are part of a much larger informational world of computer software that we are only just beginning to name and comprehend. And this is hardly surprising. For most of us, computer software as lines of code is invisible in that it is more or less unconsciously assimilated into everyday practice and behaviour as a kind of technological unconscious (Clough, 2000). It is clear that mechanical writing has now proliferated into a 'linguistic'-cum-'logistic' ecology, made up of many different kinds of programming languages, albeit chiefly sharing a common background. It is impossible to note all of these languages and what they do, so I will concentrate on just one of the most influential – Perl (Practical Evaluation and Report Language).

Created in 1987 in Santa Monica by Larry Wall, Perl is now known as the 'duct tape of the internet'. Made visible by the familiar logo of the camel, signifying an entity often thought to have been made by a committee – but which still works – Perl is a kind of mechanical writing culture with something like a million users (Moody, 2001). Built using open source methods and so standardized through many locations, Perl does not operate according to the strict logics necessary to write earlier generation programming languages. Rather, Wall, a linguistics expert, created it to mimic 'expressive' written languages on the principle of 'there's more than one way to do it'. Perl is, in other words, a language which allows a large amount of creative expression, on the principle that 'easy things should be easy and hard things should be possible' (Wall et al., 2000: 4). Thus Perl is not a minimalistic computer language. It has the capacity to be fuzzy and to migrate. And, at least in part, this was because of the openness of the development process: 'if you have a problem that the Perl community can't fix, you have the ultimate backstop: the source code itself. The Perl community is not in the business of renting you their trade secrets in the guise of upgrades' (Wall et al., 2000: xvii).

In turn, programming languages like Perl, especially in the form of various software packages and other forms of embedded system, now inhabit many of the times and spaces we take for granted. Of this point there can be no doubt. For example, the large numbers of audits carried out to test for the presence of the millennium bug show the way in which the world is now increasingly written by these languages. And the fact is that they are now so pervasive that it has turned out to be impossible to audit all of them: as the British Audit Commission (1998: 11) found out, 'some systems may be extremely difficult to locate and test'.

But there is still the problem of describing how these languages have effectivity. One answer may be to consider them as a series of writing acts, rather like speech acts, which have an 'heuretic', rather than an analytical dimension (Ulmer, 1989), based upon the inventive rather than the analytic in which language is both message and medium. Thus it becomes something of a moot point whether this means that software – as a non-representational form of action – does not rely on the activity of thinking for its ontogenesis (Hansen, 2000: 19) or whether it is simply another kind of distributed thinking of the kind routinely described in studies of distributed cognition (e.g. Hutchins, 1995). Another way to describe software may be as the delegation of yet more human functions into 'the automatic, autonomous and auto-mobile processes of the machine' (Johnson, 1999: 122), as part of a process of externalization and extension of the vital based, for example, on an apprehension that the human body is simply 'too slow' to cope with many of the situations of contemporary life. Whatever the case, life is, so to speak, increasingly re-inscribed as what Clark (2001) calls 'wideware', a whole set of different cognitive technologies working to act.

These kinds of problems of description are only compounded by the drift in many computer languages from being simply imperative to inferential and from machinic to quasi-biological emergent architectures which allow programs to adapt and grow (Johnson, 2001). For, as programs become more open, expressive and adaptive so they start, sometimes on purpose, to mimic biological processes. For example, a number of computational algorithms now operate on explicitly genetic principles (or other similar biological analogies) and these algorithms can be found operating in everyday situations (for exam-ple, medical diagnostic systems, credit-scoring systems, traffic management systems, call centre routing systems, and so on) (Thrift and French, 2002).

The mention of the biological brings us to the final kind of mechanical writing that now saturates contemporary timespaces, namely that to be found in the realm of genetics as the writing of the code of life itself. The new techniques of genetic sequencing developed in the 1970s were often seen as equivalent to producing new forms of writing, 'to be fluent in a language one needs to be able to read, to write, to copy, and to edit in that language. The actual equivalents of each of these aspects of fluency have now been embodied in technologies to deal with the language of DNA' (Jackson, cited in Kay, 2000: 1). Whether this was an accurate analogy is a moot point, of course, as Kay (2000: 2) has famously pointed out. Be that as it may, it may be that genetics *is* now becoming a form of writing because of the intervention of the computer program (Doyle, 1997). As bioinformatics and biocomputation grow in scientific stature, so the program makes its way into biology and biology makes its way into the program, producing strange new hybrid forms of the vital and the programmatic in which the operators of writing produce new domains.

But the new complexes of normalization do not stop at mechanical writing. The new normalization is increasingly felt in *the body*. I am quite sure that one of the key battlegrounds of standardization in the early twenty-first century will continue to be the attempt to invent a new kind of body which, unlike the regulated industrial bodies of much of the twentieth century can be flexible, performative and, above all, inventive. Producing flexible, performative and inventive bodies relies upon the generation of affect. Whether we think of affect in Deleuzian terms as a general plane of sensation, or in Jamesian terms as the somatic feedback to the brain (as, for example, when an infant is seized by her or his own crying) or in Tomkins' terms, as a bridge between the biological and the cultural by dint of the fact that the feeling subject recurs in others as well as in its own further response to this feedback (as in the feeling of shame), it seems clear that increasingly modern societies work by engineering affect (Katz, 1999, Sedgwick and Frank, 1995).

Working especially via the media, which amplifies affect through selected forms of gesturality (Agamben, 1998), and, at the same time, a greater and greater appreciation of the maintenance of bodily interaction (in part a function of media technologies that allow such interaction to be studied), the engineering of affect has become common in several social domains: the state and politics (Nolan, 1998), business (Thrift, 2000) and, increasingly, civil society (for ex-ample, through education).

This engineering of affect is clearly being used to produce new forms of capitalist subjectivity. In business bodies are being reconfigured using insights from a wide range of different sources of expertise. For example, various sources – psychology, phenomenology, Buddhism, New Age – are being used to produce technologies that can reconfigure the management body as creative and inspirational by working on affect. Three of these corporeal technologies be-came key in the 1990s (Thrift, 2001). The first of these was *organizational*, and consisted of technologies that brought bodies into creative alignment. In part-icular, optimal alignment was considered to occur through the use of teams and projects.

However efficacious they may or may not be, the fact is that teams and projects are now regarded as the main way in which bodies can be aligned to produce creativity. The intention is to produce concerted blocs of time in which people can come together productively to push through a particular creative project. In turn, all over the world, offices are being redesigned to cope with this way of working. 'Hives' and 'cells' are being replaced by 'club' and 'den' environments (Duffy, 1997).

The second technology was *inspirational* and consisted of the careful design of events which would enable organizations to interact creatively on a larger scale. A whole series of these technologies now exist, from conferences and seminars, through to courses and workshops. Their purpose is in part to disseminate information but it is also in part to keep the current of inspiration going. Many of these events are minutely plotted and the smaller of them use a number of summative body techniques, from performance (e.g. theatre, dance, opera), through body control techniques like Aikido or the Alexander Technique, to various forms of ritual (especially of the New Age variety).

The final technology to emerge was *ideological*. Each organization has to have narratives which will sustain it, especially in circumstances in which there might be constant jumping between projects, in which the organization was likely to be dispersed over many locations, and in which there might be high personnel turnover. Thus the vogue for corporate story-telling, corporate websites, and the like, and, not coincidentally, a vast outpouring of business books, magazines and television series, each of them telling exemplary stories of what it is like to be 'manager'.

In each of these cases what we see are standardized formats intended to change the management body by changing space and time. From the vagaries of the modern office, through the controlled otherness of the event, through to new iconological formats, what we see are attempts to change the background of timespace by changing the way the body lives (Thrift, 2000). Change time-space, change bodily stance to the world.

In turn, however, there is the problem of how to measure and audit these reconfigurations. It is no accident, I think, that there is an outpouring of work currently which is trying to stabilize this knowledge through a whole series of techniques of measurement. Most of these techniques no doubt may currently seem a little strange but in twenty years time there is no reason to think that this will be the case. For example, much effort is currently being expended on measuring consumer attention by developing 'attentionscape' technologies, especially using small headsets, on the grounds that 'in post-industrial societies, attention has become a more valuable currency than the kind you store in bank accounts' (Davenport and Beck, 2001: 3). The age of the 'engagement index' may not be far off.

Then, I want to consider one other complex of normalization which is associated with *built form*. It is hardly a remarkable statement to argue that built form is a key element of social construction. Indeed, the recent work in the sociology of science on the laboratory and other similar buildings (e.g. Galison and Thompson, 2000), yet alone the use of buildings in institutions like ritual (such as medieval cathedrals), makes clear. But until recently I do not think it is unfair to say that buildings have tended to be regarded as, generally speaking, objects with few powers of their own. Now, however, this is no longer the case. We can see this theoretically in, for example, the attention given to Foucault's work on the rigidly controlled spaces of the Panopticon or Mettray. More generally, it is manifested in the general interaction between theory and architecture (to the point where it is often difficult to separate theoretical expression of architecture from the architectural expression of theory), for example, in the work of Bernard Tschumi, or Steven Holl, or Lebbeus Woods, or Greg Lynn, and in the interaction between performance and architecture, as in the work of numerous stage designers and choreographers. Then, practically, there is an industry which ranges all the way from the discovery of new kinds of lighting effect which can produce new senses of space, through the growth of building interior design, to new locations in the division of labour like 'Facilities Manager' which are intended to produce a fuller understanding and more active management of buildings, all of which depend on the production of new spatial and temporal knowledges. These moves have been made possible against a general background of economic calculation that has become increasingly sophitcated about how buildings can be valued: from securitization of actual properties to the continuous discourse about how buildings can be designed so that the people in them will work more or better.

In particular, buildings are increasingly seen as powerful means of moulding subjectivity through their ability to normalize the range of bodily responses that is possible in a space. But this does not mean that buildings are necessarily seen as 'walk-through-machines' in which bodies can only play a set part. They can also be seen as a means of extending the range of responses of which the body is capable by producing enhanced qualities of interaction.

Certainly, this is the reasoning behind much of the work on the modern office. Here the idea is clearly to understand buildings as evolutionary entities able to adapt to circumstance, which can produce management and other bodies that are similarly evolutionary and able to adapt to circumstance. The move to 'den' and 'club' from 'hive' and 'cell' environments is specifically heralded as a move to a new spatial vocabulary which can produce better business performance by cultivating a shared 'business intelligence'. It is meant to produce high occupancy spaces which are complex and also manipulable, allowing the work process to be constantly redesigned and a complete spectrum of skills to be accommodated. Most importantly, these offices are meant to allow sharing and interaction which will enable creativity and innovation to blossom. The management body, in other words, is placed within a new set of settings which extend its range and possibilities (Duffy, 1997). What is part-icularly interesting is that noone can be sure that these buildings meet their objectives, in the absence of any clear measurement strategies, and yet there seems to be little general scepticism about their effects.

THICKENING CAPITALIST RELATIONS

Let me now turn to capitalism's attempts to stabilize the changes in timespace that it has in large part created. Very often nowadays, capitalism (for which, usually, read large corporations) is depicted as a vast circulating behemoth.[2] On the other hand, anyone who has ever worked in a large corporation would have to greet these kinds of leviathan accounts with a certain degree of scepticism. Such corporations are often loose coalitions of competing or even opposed interests, ill thought-out plans, and grudging implementations held together by inspired acts of improvisation. And things really do fall apart on a fairly regular basis – and not just because of market downswings.

So how is it that large corporations which, for all their resources, so often seem such hesitant and fragile enterprises, are able to take hold of and even extend their grip over a continuously modulated world of multiplicities? I want to argue that it has been concerted work on the simple mechanics of capitalism, especially though the formation of new spatial and temporal knowledge, which has led the extension of capitalist social relations to their current dominance, work which has produced a world in which things still go wrong on a regular basis, but do less harm and can be put right much more quickly than before. Capitalism, in other words, has built up a *resilience* which now makes its existence so stickily pervasive when compared with so many previous social systems. I will focus on three different circuits of spatial and temporal knowl-edge, each of which has come into existence since the 1960s, in order to make my point.

The first of these I have called elsewhere the 'cultural circuit' of capital (Thrift, 1997; 1998; 2000). Here I would describe that circuit as one of contin-uous critique of extant business practice (Boltanski and Chiapello, 1999). The three chief producers of this critique are business schools, management consultants and management gurus. Business schools were first founded in the late nineteenth century and early twentieth in the United States. However, save for a small elite, the main phase of expansion in the United States took place much later, from the 1940s on, on the back of the MBA degree. In the rest of the world, business schools were slow to be founded, but in the 1950s and 1960s they began to open and expand in Europe and subsequently in Asia. Nowadays business schools are the jewels in the crown of a vast global executive education market (Thrift, 1998).

Management consultants also date from the later nineteenth and early twentieth centuries. Often described as unacknowledged legislators, manage-ment consultants offer advice to business on such a large scale that a case could be made that they have simply become an extension of firms. Whatever the case, it is clear that they are important producers and disseminators of business knowledge, able to take up ideas and translate them into practice and to feed practice back in to ideas (Micklethwait and Wooldridge, 1996; Clark and Fincham, 2001; Vann and Bowker, 2001).

Finally, management gurus are chiefly a phenomenon of the later twentieth century, consisting of various well-known academics, consultants and business managers who have been able to package their ideas as aspects of themselves. Though there is a clear genealogy, modern management gurus are usually dated from Peters and Waterman's *In Search of Excellence*, published in 1982. Gurus tend to develop formulaic approaches to management, which must play down context for the sake of rhetorical force.

Producers of business knowledge necessarily have a voracious appetite for new knowledge which can continue to feed the machine of which they are a part. So they do not just produce knowledge from within. They are also constantly on the hunt for knowledge from without which can be adapted and brought within. Thus almost every aspect of human knowledge is available for incorporating (Vann and Bowker, 2001; Thrift, 1999).

These producers are responsible for a range of different kinds of business knowledge. Put schematically, it is possible to say that this knowledge has three main functions. The first is the provision of general principles of business life – 'do this, don't do that' – a kind of grammar of business imperatives. The second is as a primer which tells managers how to attain particular goals. The third is an intelligence-gathering function – concerned with how business practices are working out. In other words, what is being produced is a process of endless, relentless, and continuous critique of the status quo and the incorporation of that process of critique via new management practices (Boltanski and Chiapello, 1999).

However, these three producers could not exist in their modern form without a symbiotic relationship with the media which both publicizes and distributes their wares. There are numerous ways in which the media intervenes. First, through the production of standard media like books, magazines, newspapers, internet sites, and television. The importance of journalists as translators of business ideas, coupled with the way in which the media provide outlets for writing for the knowledge producers to display their wares, are underlined by these media (Furusten, 1999). A second element is the increasing scale of specialized business media. These range all the way from general and industry-specific magazines to the new breed of consultancy-sponsored magazine which emulates *The Harvard Business Review*. The model provided by *Fast Company*, first published in 1995, has also proved particularly influential, leading to a large number of copycat magazine formats (Thrift, 2001). A third element is the growth of media intermediaries – press officers, publicity consultants, design consultants, advertising agencies, and so on – which have become more important as business ideas have increasingly come to resemble brands. Then, a fourth element has been the re-engineering of the face-to-face meeting through the continual production of conferences, seminars, workshops, and the like. These events serve both as disseminators of new business practices and as motivational fuel.

I call the second circuit the circuit of service and repair. Service and repair is constantly and consistently underestimated as a set of knowledges. It is almost always seen as a secondary activity and this perhaps explains its neglect. And yet even a glance at Yellow Pages and the lists of service engineers and repairers shows how absolutely crucial this activity is. But what is interesting is that, over the last twenty or thirty years, service and repair has been carefully reworked as a customer service industry which now functions on a massive scale (Rifkin, 2000). A good example can be found in the contemporary information technology industry whose core consists of machines which are still remarkably unreliable (Norman, 1999) and need constant service and repair. Here all manner of service organizations have grown up which involve not just knowledge of the product but also how to manipulate space and time to produce fast repair of that product, from emergency backup server farms to companies like the EMC Corporation which is involved in both data storage products and service and repair.

I call the third circuit the circuit of logistical expertise. Again, over the last twenty or thirty years, logistics has become a specialist set of practical and external knowledges with a vast array of its own MBA courses based upon the increasing importance of speed in the product cycles of corporations (as, for example, through the rise of merchant subcontracting), the rise of technologies that enable locational attributes to be systematically assessed in real time (as in recent work on geographical information systems), and, of course, the importance of driving down inventory.

CONCLUSION

In this chapter, I have tried to produce a sense of a Euro-American world hoving into view in which a new kind of transcendental empiricism holds sway, one which is both created by and is producing new spatial and temporal assemblings and arrangings. In some senses, we might count this as a more open world, in that these assemblings and arrangings are not attempts to close down possibility but rather to modulate it – 'might' is right. But, increasingly, that 'might' is also what nowadays produces might.

In other words, as I have shown, this does not mean that this is a world without normalization. Rather, the form that normalization takes has shifted. Because of a set of socio-technical changes that I have outlined, rules are increasingly able to be context-specific ('location-aware') and so adaptive. We might say that the continuous activity of normalization has taken a kind of Aristotelian turn – towards rules that are based on stances to judgement that are enabled by much greater knowledge of particular spaces and times than would have been possible in the past. So, increasingly, through the enactment of these new timespaces, global standards will also be local, and vice versa. How rules are will change. This shift has benefits, in that it opens up all kinds of new possibilities; such rampant overcoding is bound to leave spaces of ambiguity and dispersal, it may even lead to new political rationalities. But it also means that the exercise of control will be much more insidious because it will be so often woven into the background of the everyday times and spaces of work and leisure; bodies will take up preconditioned stances too soon, and the flicker of mental life will be consequently all the harder to maintain. No evisceration, then, but certainly a need for greater and more subtle political effort.

ACKNOWLEDGEMENT

This chapter has benefited from being read at two conferences, the conference on New Spacings and Timings held at Palermo, and the Third Graz Bienniale of

Architecture and Media. I thank the audiences at these conferences for their comments. Particular thanks go to Bruno Latour, my discussant at the first conference and Lars Lerup, my discussant at the second conference. John Urry gave the chapter a careful and inspiring going-over which has much improved its clarity and logic. Henry Yeung and Jamie Peck also provided some important correctives. The National University of Singapore kindly gave me the opportunity of a Distinguished Visiting Professorship to work this chapter up for publication.

NOTES

1   This proliferation is therefore often concerned with possibility arising from combination, as much as virtuality arising from novelty in Deleuze's terms.

2   Most recently, for example, in a very interesting book, Hardt and Negri (2001) have presented a modern variant of this kind of depiction which seems to cede capitalism an almost Illuminatus-like capacity to conspire and compel, albeit in a dispersed and deterritorialized way, occasioned especially by its fiscal and communicative capacities.

Chapter 9

BEYOND THE CLUSTER – LOCAL MILIEUS AND GLOBAL CONNECTIONS

Anders Malmberg

INTRODUCTION

Economic geography in a knowledge-based economy

How is the performance of firms affected by the conditions that prevail at their place of location and why do certain local milieus prosper while others don't? Why do similar or related firms so often locate nearby each other and how are such patterns of regional specialization reproduced over time? The increased focus in recent years on the role of learning and innovation as key processes behind sustained industrial competitiveness has brought some of these core considerations of economic geography to the research agenda in a wider community of economic scholars (Storper, 1995; Porter, 1998; Fujita et al., 1999). These questions are also being addressed in a new way. A contemporary view of these issues can be summarized in four assertions.

First, in today's knowledge-based economy, the ability to innovate is more important than cost efficiency in determining the long-term ability of firms to prosper. Innovation is here defined broadly as the ability to come up with new and better ways of organizing the production and marketing of new and better products (Porter, 1990; Lundvall, 1992; Nelson, 1993; Nonaka, 1994; Grant, 1996). This does not mean that cost considerations are unimportant, but simply that the combined forces of globalization of markets and deepening divisions of labour make it increasingly difficult to base a competitive position on cost-advantage only.

Second, innovations predominantly occur as a result of interactions between various actors rather than as a result of the solitary genius (von Hippel, 1988; Håkansson, 1987; Lundvall, 1992). This fits with a Schumpeterian view of innovations as new combinations of already existing knowledge, ideas and artefacts (Schumpeter, 1934). Most innovations are based on some form of problem solving. Someone perceives a problem and turns to someone else for help and advice. In an industrial context, these interactions often follow the value chain (Malmberg and Power, 2002). A firm facing a particular problem turns to a supplier, a customer, a competitor or some other related actor to get help in specifying the problem and defining the terms for its solution. From this follows that the level of analysis for understanding the processes of industrial innovation and change is some notion of an industrial system or network of actors carrying out similar and related economic activity.

Third, and this is where 'geography' enters the picture. There are a number of reasons why interactive learning and innovation processes are not space-less or universal, but on the contrary unfold in a way where geographical space plays an active role. Spatial proximity carries with it, among other things, the potential for intensified face-to-face interaction, short cognitive distance, common language, trustful relations between various actors, easy observation and immediate comparison (Malmberg and Maskell, 2002; cf. Amin, in this volume). In short, spatial proximity seems to enhance processes of interactive learning and innovation and therefore industrial systems should be assumed to have a distinctly localized component.

Fourth and finally, an implication is that the knowledge structures of a given geographical territory are more important than other characteristics (such as general factor supply, production costs,etc.) when it comes to determining where we should expect economic growth and prosperity in today's world economy.

This chapter is structured into three main sections. The next section is about the merits and shortcomings of one of the more influential approaches in economic geography in recent years – the cluster approach. The subsequent section offers some specific proposals on how a cluster approach could and should be applied in order to enrich empirical research. The final section revolves respectively around the characteristics and the role of local interaction and global connections, and proposes a somewhat revised research agenda for future analysis of the impact of geographical space on industrial performance.

CLUSTERS AND CLUSTERING

Since the publication of Michael Porter's (1990) book, the cluster concept has become widely circulated in academic as well as in policy circles. In a recent paper, Martin and Sunley (2001) scrutinize this cluster trend in economic geography and related disciplines and advance a number of more-or-less justified points of critique. Seen as a review of a rapidly proliferation research field, the paper is a great source of information. In my view, however, their critique in some important ways overshoots the target and fails to recognize some of the genuine advancements that 'the cluster turn' has actually brought with it.

The contribution of the cluster approach

The cluster approach, as presented by Porter (1990; 1994; 1998; 2000) and subsequently developed by him, his associates and others (Enright, 1998; Malmberg et al., 1996; Malmberg and Maskell, 2002), brings some genuine contributions to the analysis of some key issues in economic geography. First, it provides a way to describe the systemic nature of an economy, i.e. how various types of industrial activity are related. Porter's starting point here is the cluster chart (see Figure 9.1). Beginning with the firms in the industry where we find the main producers of the primary goods of the cluster (be they heavy trucks, telecom equipment or popular music), the chart proposes a way to analyse how these firms and industries are connected to supplier firms and industries providing various types of specialized input, technology and machinery and associated services, as well as to customer industries and, more indirectly, related industries.

Figure 9.1   The cluster chart: actors in an industrial cluster

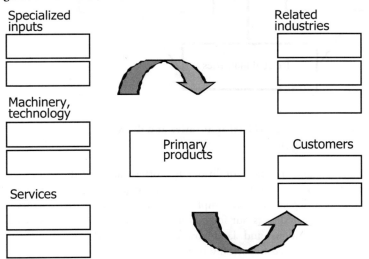

This way of approaching the 'systemicness' of economic activity has several advantages. It broadens the scope for analysing interactions and inter-dependencies between firms and industries across a wide spectrum of economic activity. Another advantage is that it bridges a number of more-or-less artificial and chaotic conceptual divides that characterize so much work in economic geography and related disciplines. These include, for example, manufacturing vs. services, high tech vs. low tech, large companies vs. SMEs, public and private activities, etc. A single cluster, defined as a functional industrial system, may embrace firms and actors and activities on both sides of each of these divides (see also Dicken and Malmberg, 2001).

Furthermore, Porter's model of the determinants of competitiveness in cluster, known under the 'brand' of the *diamond model*, identifies a number of mechanisms proposed to foster industrial dynamism, innovations and long-term growth (see Figure 9.2). Essentially, the model is built around four sets of intertwined forces respectively related to factor conditions, demand conditions, related and supported industries, and firm structure, strategy and rivalry. The point here, which for example Martin and Sunley (2001) fail to acknowledge in their recent critique, is that the treatment of these factors include several points that are indeed novel.

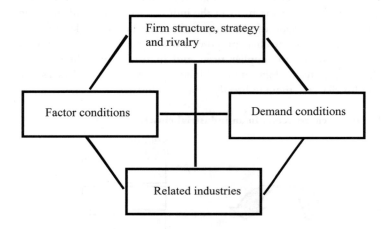

Figure 9.2 Forces that make a cluster innovative and dynamic: Porter's diamond

First, the treatment of factor conditions in Porter's account does offer some rather radical departures from the prevalent view, e.g. in pre-1990 economic geography. One such departure is the emphasis on the role specialized factors and factor upgrading, which redirects our focus from the classical notion of availability and cost for capital, labour and land towards the type of specialized factor conditions that are developed historically to fit the needs of a particular economic activity. These are important as factors of location since they are difficult to move and difficult to imitate in other regions (cf. Maskell et al., 1998). Another, perhaps more original, idea is that of selective factor disadvantages as a factor promoting dynamism and long-term growth. Arguably, no previous account had explicitly made the point that shortcomings in factor conditions (such as labour shortage, and high wages, scarce natural resources, expensive electricity, etc.) can actually trigger technological and institutional innovations that will in the longer term contribute to the competitive success of the firms in a specific location.

Second, the treatment of the demand side as a primarily qualitative factor is original. Most previous models have emphasized access to a large market as an important location advantage. Porter's account, in contrast, alerts us to the fact that it is the sophistication of demand that matters – if we are interested in inno-vation and long-term competitiveness. According to this view, the locationally advantaged firm is the one that is in a position to receive and react to signals of sophisticated demand, rather than simply one that is blessed with 'many customers' in the local market. This idea is also present in other recent appro-aches to the dynamics of industrial systems. In Eliasson's (2000) notion of the competence bloc, the 'competent customer' plays a key role.

Third, the importance of local rivalry is made much more explicit than in previous models of spatial agglomeration. That a firm may gain advantages from being located close to other firms in the same industry is of course a key insight in classical agglomeration theory. Rarely, though, has this advantage been attributed to the fact that spatial proximity between rivals will trigger dynamism and growth. The idea is that local rivalry adds intensity and an emotional dimension to competition that most firms perceive in the global market. The firm down the road is often seen as the 'prime enemy', a bit like the rivalry between neighbouring football clubs (AC Milan vs. Inter Milan or – to use an example that Peter Dicken would immediately recognize since he has for so long suffered emotionally from supporting the losing side – Manchester City vs. Manchester United). Firms in a local milieu tend to develop relations of rivalry where benchmarking in relation to the neighbours is more direct, partly for reasons of local prestige and partly because direct comparison is simplified. One could speculate that there are at least two reasons for the latter. First, it is easier to monitor the performance of a neighbouring firm than a competitor far away. Second, if one firm displays superior performance, it is obvious to everyone that this cannot be 'blamed' on different external conditions, since they are in principle identical for all firms in the local milieu (cf. Malmberg and Maskell, 2002).

On these points, at least, it should be acknowledged that the cluster approach, as developed by Porter and others, has contributed to genuine pro-gress. The role of specialized production factors and selective factor disadvant-ages, sophisticated customer demand and local rivalry are novel and innovative proposals that have enriched our understanding of why conditions in a local milieu in general, and agglomeration of similar and related firms in particular, might promote superior firm performance.

Conceptual confusion

At the same time, it is easy to agree that there is a good deal of conceptual fuzziness surrounding the cluster issue (Markusen, 1999; Martin and Sunley, 2001). In my view, some of it is rather basic and should be resolved before turning to 'confusions at a higher level'. The really disturbing lack of clarity exists as a matter of fact at the most basic level: what is meant by the terms 'cluster' and 'clustering'? This seemingly trivial question is causing continuing and increasing problems. I am not thinking here of subtle definitional issues related to the scales, boundaries of and criteria for identification of clusters. Instead, in my view, the main confusion is related to whether clusters and clustering are primarily *functional* or indeed *spatial* phenomena. On this particular issue, Porter himself has contributed to the conceptual mess by presenting quite different basic definitions in various texts since 1990. Compare, for example, the following few quotations. In his 1990 book, Porter writes:

> The competitive industries in a nation will not be evenly distributed across the economy ... A nation's successful *industries* are usually *linked* through *vertical* (buyer/supplier) or *horizontal* (common customers, technology, channels, etc.) *relationships* ... The reasons for *clustering* grow directly out of the determinants of national advantage and are a *manifestation* of their *systemic character. One competitive industry helps to create another in a mutually reinforcing process.* (Porter, 1990: 148–9; emphasis added)

Only after having thus asserted that clusters are sets of *functionally* interrelated industries (within the spatial context of a nation) does Porter go on to discuss the role of geographic concentration:

> Geographic concentration of firms in internationally successful industries often occurs because the influence of the individual *determinants* in the 'diamond' and their mutual reinforcement are *heightened by close geog-raphic proximity* within a nation. (Porter, 1990: 156–7; emphasis added)

Thus in the 1990 book, it is obvious that Porter regards clusters *as functionally related industries*, while at the same time observing that such functional clusters 'often' seem to be prone to 'geographic concentration' because spatial proximity amplifies the mechanisms that make clusters of industries dynamic and innovative. Then, throughout the 1990s, Porter adopted a view according to which geographic concentration gradually becomes an integral part of defining a cluster. Thus, in a recent paper, Porter (2000) writes:

> *A cluster is a geographically proximate group of interconnected companies* and associated institutions in a particular field, linked by commonalities and complementarities. The geographic scope of the cluster can range from a single city or state to a country or even a group of neighbouring countries. (Porter, 2000: 254; emphasis added)

Now clusters are defined by *geographical proximity*, even though the scale of this geographic concentration is kept very elastic (from the single city to a group of neighbouring countries). This gradual slide in the definition of the cluster concept is unfortunate. As a matter of fact, I will argue below that it is a main source of confusion.

The conceptual confusion has been aggravated by the impact of 'cluster thinking' in the policy arena. When turned into policy, the cluster concept has taken on a third meaning, partly separate from the two dimensions discussed above. In policy circles, a cluster has become more-or-less synonymous with a policy programme and a number of more-or-less concerted policy actions. Here, a cluster is seen to come into existence when an actor (often situated in a public institution rather than in a private company) identifies a cluster, whether exist-ing or 'dormant' (or 'potential', or 'emerging'), gives it a name (often ending with 'Valley') and starts acting in order to develop it consciously (Rosenfeld, 1997; Raines, 2001). This could be seen as a *discursive* definition of the cluster concept and it might have a resemblance with either the functional or the geog-raphical dimension already discussed. The broader issue of whether the cluster models form an appropriate base for industrial, innovation and/or a regional policy is interesting and controversial in its own right, but has to be left out of this chapter for reasons of limited space.

Getting the semantics right: industry clusters and spatial clusters

The above has implications for how economic geographers and others should approach the clustering issue. First, it would indeed be practical if we could collectively strive to establish a terminology that is as free as possible from basic confusion. It is deeply unsatisfactory to develop a scholarly conversation around a core concept – the cluster – the meaning of which various participants in the conversation have different opinions, not in detail but at the level of basic definitions. We certainly need one concept to notate the idea of functionally linked economic activity. The 'industrial system' would seem to be an appropri-ate generic term, but since the 'C-word' is presumably going to be around for a while, 'industry cluster' could be a useful alternative. When, on the other hand, we face geographical concentrations of similar or related economic activity, we could preferably use the traditional term agglomeration, or possibly 'spatial (or localized) cluster' in order to avoid some of the confusion.

But this is not simply a question of terminology. It seems obvious that (functional) industry clusters will normally not be confined to, or contained within, any narrowly defined and spatially bounded scale. On the contrary, most industry clusters will have widespread global connections and if we could be able to identify their boundaries in spatial terms, the spatial scale would in most cases certainly not be an urban region. By making spatial configuration (i.e. degrees of agglomeration) an attribute of an industry cluster rather than part of its definition, I believe we could establish a platform for more fruitful analyses of how 'geography' comes into play in the overall process of industrial compe-titiveness, growth and transformation. In other words, rather than trying to squeeze 'cluster charts' into narrowly defined regions (where they rarely will fit in), we should research the hypotheses found in the 'diamond' regarding the role of proximity and local milieu on the proposed mechanisms leading to competitiveness.

RESEARCHING CLUSTER DYNAMICS

A preoccupation with spatial readings of the cluster concept has contributed to sidetracking empirical research on clustering. The introduction of the cluster concept could have triggered lots of research on the fruitful issue of how industrial transformation occurs as a result of interactions within and across industrial systems (i.e. clusters defined in the functional sense) and the role of geographical proximity (concentration or agglomeration, i.e. clustering in the spatial sense) in such processes. Instead, I would argue, there has been far too much focus on interaction between firms within geographically defined spaces and numerous rather pointless attempts of trying to assess the degree to which there is actual interaction going on locally and thus whether a specific region can indeed be said to contain a 'fully-fledged' or a 'true' cluster (cf. Martin and Sunley, 2001).

Partly misconceived hypotheses ...

Thus, the empirical validation of the propositions advanced in the cluster literature leaves a lot to be desired. First, there has been a general reluctance to spell out the theoretical propositions made in a form that would make it possible to subject them to systematic empirical validation. Consider the following proposals, for example:

1   Firms that meet sophisticated demand from customers in the local milieu will be forced to innovate at a higher pace than other firms.

2   Firms that collaborate more on technology with firms and other actors (e.g. universities) in the local milieu will innovate more.

3   Rivalry between similar firms in a local milieu will be more intense, almost emotional, and this will create pressure to innovate in order to outsmart the local rival. Therefore, firms with nearby rivals will be more innovative than firms that have their main competitors located elsewhere.

4   Knowledge diffusion will be more rapid among local firms than among globally dispersed firms through informal interaction in the local milieu as well as through flows of people in the local labour market.

These, I would argue, are all interesting and researchable hypotheses that could be deduced from the cluster literature, based on the underlying argument that the forces that enhance the dynamism of an industry cluster are strengthened by geographical proximity via a series of mechanisms. The typical outline of a study approaching these issues, however, is to start from a notion of a spatially defined cluster – i.e. a regional economy where there exist a number of successful firms within one or a few interlinked industries – with the underlying assumption that if this is indeed a 'true cluster', we should expect most key relations between firm to take place within this spatially bounded system.

... lead to disappointing results

Therefore, it is not surprising that, to the degree that empirical analyses have actually tried to catch these issues, the results are mainly rather disappointing. This is not the place for a thorough review of empirical studies (see Malmberg and Maskell, 2002; Martin and Sunley, 2001 for some such overviews). I believe, however, that the following summary of empirical results, despite being somewhat bold and stylized, is fairly accurate:

1   There are generally limited transactions between firms in the local milieu. When asked questions about where the most important suppliers or customers are located, most firms report on spatially fairly extended networks. Global connections tend to dominate over local networks (e.g. Markgren, 2001; Larsson, 1998; Larsson and Lundmark, 1991; Angel and Engstrom, 1995).

2   There is limited formal collaboration between firms (or between firms and other organizations) locally. Such collaborations tend to follow the value chain and therefore to be fairly globally extended (Larsson, 1998; Mackinnon et al., 2003; Fuellhart, 1999; Owen-Smith and Powell, 2002).

3   There is indeed sometimes intense local rivalry. On the other hand, many firms report that they have few direct competitors worldwide and that those few rival firms are located elsewhere (Malmberg et al., 2000; Glaeser et al., 1992; Audretsch and Feldman, 1996; Baptista and Swann, 1996; 1998; Larsson, 1998).

4   Firms in a local milieu have rather good knowledge about each other, even though they often do not know where this knowledge comes from and whether it is valuable. There tends to be labour mobility between firms locally, but many firms see this more as a problem than as an advantage (Lawson, 1999; Almedia and Kogut, 1999; Dahl, 2002).

A crude summary of the above is that there tend to be modest commercial relations between firms within spatial clusters. Other types of collaboration are more common locally, but such relations normally extend well beyond the borders of narrowly defined regions. The degree of local rivalry varies; informal knowledge exchanges do occur and local labour mobility is presumably an imp-ortant factor.

These results have led to a gradual shift in research focus. An initial preoccupation with analyses of transaction links between firms in local milieus gave way during the 1990s to an increased emphasis on other forms of inter-firm collaboration, such as joint projects in technological development. These studies too produced mainly disappointing results in the sense that it often turns out that the most innovative firms are indeed globally well connected. This, in turn, led many scholars in the field to give increased attention to the more informal, subtle and often almost unintended interactions taking place as result of the predominantly localized nature of everyday life (unplanned meetings in bars and restaurants, gossip and rumour, spontaneous monitoring of competitors nearby, etc.). In the terminology of Storper (1995; 1997), analyses of traded interdependencies among firms in a local milieu have gradually given way to studies of untraded interdependencies. Much work of the latter type is certainly needed. The issue of 'learning by being there' (cf. Gertler, 1995; in this volume) deserves further attention, and there is scope for further analyses of the intricate webs of relations and interactions – between people as much as between firms – that take place in the local milieu (see further below).

The underlying problem, though, is the expectation at the outset that there should be a high magnitude of local inter-firm relations for a cluster to be said to exist – and indeed for a cluster to be dynamic and prosper. In that case, the disappointment is largely a result of an initial misconception. If we accept that industry clusters are normally not confined to local milieus, then we should not expect them to be primarily locally integrated. There would presumably be much to gain from dropping the underlying assumption that 'the more localized interaction, the better'.

Beware of introvert clusters

Interestingly, some clear pointers in this direction are found in Porter's original account. In a section on 'insular clusters', Porter (1990: 171–2) writes about how complacency and an inward focus often explain why nations lose competitive advantage when firms face too little pressure and challenge to upgrade constantly technologies and processes. When this happens, firms tend to hesitate to employ global strategies to offset local factor disadvantages or to tap selectively into advantages available in other places:

> The cluster itself, particularly if it is geographically concentrated, may contain the seeds of its own demise. If rivalry ebbs and home buyers become pliant or lose sophistication, there is a tendency for the *local cluster* to *become insular*, a closed and inward-locking system. The problem is exacerbated *if* most firms lack significant international activ-ities and their *primary commercial relationships are with each other* (for example suppliers sell almost excl-usively to a single domestic industry). *Firms, customers and suppliers talk only to each other. None brings fresh perspectives.* (Porter, 1990: 171; emphasis added)

Such clusters are seen to be vulnerable to structural change and to chance events. Moreover, Porter points to the inherent risk of lock-in that follows from strong local specialization. When skills, assets and strategies are highly specialized to a particular industry structure, firms can adapt incrementally but may have difficulties dealing with radical technology shifts (cf. Asheim, 1996; Grabher, 1993; Maskell and Malmberg, 1999). So, in a sense, the idea that clusters are indeed spatially defined entities and that a dynamic cluster is characterized to be intense local linkage and interaction does not really originate in Porter's analysis; it is more a result of the interpretation by others.

A REVISED AGENDA:
UNDERSTANDING LOCAL INTERACTIONS AND GLOBAL CONNECTIONS

Still, there are various reasons why proximity and local milieus are important in the process of the overall transformation both of industrial systems ('industry clusters') and specialized local and regional economies ('spatial clusters'). In this final section, I will discuss this further by addressing three specific issues, which I believe have been neglected in research on clustering: the role of large global firms, the role of local labour markets, and the spatiality of inter-personal – rather than inter-firm – interaction in learning and innovation processes.

What about global firms in spatial clusters?

A key point of departure in much contemporary writings is that 'the local', despite claims of an ever-broadening and ever-deepening process of global-ization, remains important. So what do we mean when we say that local things have become more important? Clearly, regarding many aspects of economic life, spatial proximity means less than it used to. Raw materials can be shipped, global markets can be penetrated from many different places and certain types of information can be easily accessed regardless of location. And there are indeed 'global firms' (see Dicken, in this volume). This is the first neglected area to bring into the discussion. Research on industrial transformation today is divided in two separate worlds. One of them focuses on regionally defined systems of firms and other institutional actors and how interactions and spillovers among these create innovation, competitiveness and prosperity. Here, the role of large global firms tends to cause unease. The 'true' actors of such milieus are locally owned small and medium-sized firms, while globally oriented transnational corporations (TNCs) one way or another are seen as alien to the idea of a dynamic local milieu. This is most explicitly expressed in some of the work on industrial districts, but the same model of thought is implicitly expressed in much work on regional clusters (cf. Malmberg and Sölvell, 2002).

In the other camp, students of international business focus on the way in which globally organized companies develop their strategies, tap into, integrate and disseminate various bodies of knowledge and restructure their operations at a much more global scale (Andersson et al., 2001; Andersson and Forsgren, 2000; Holm and Pedersen, 2000). Far too much are these two worlds held apart (see Amin and Cohendet, 1999; Amin, in this volume). Great potential exists, therefore, in integrating these two points of departure, and I believe one necessary condition for bridging this gap is to drop the rather rigid 'local' focus in much contemporary cluster research.

The case for a labour market approach to spatial clustering

But let us return to the question of what is 'local'. One basic and obvious point here is: people are. While goods, money and certain types of information can indeed travel the world with little friction, people can't. Therefore, it would seem obvious that analyses of localized phenomena like spatial agglomeration of similar and related economic activity (i.e. regional clustering in the terminology proposed here) should indeed focus more on the factors that are related to the restricted mobility (or even spatial fixity) of people.

Bringing people into cluster research means, among other things, allowing more explicitly for the functioning of labour markets and the role of human skills (and everything that leads up to it, including educational systems, housing markets, service facilities, health care, insurance and pension systems, rules and norms, etc.) to occupy a more central position in the analytical framework. In one sense, this has already happened with the increased focus on the role of trustful relations in interactive learning process and the spontaneous knowledge and information spill-over that seems to occur in predominantly local milieus (cf. Gertler, 2002c; Maskell et al., 1998; Malmberg and Maskell, 2002). Still largely missing, to my knowledge, are analyses of the functioning of local labour markets in relation to the spatial clustering phenomenon.

This was indeed part of Marshall's (1890) original account of the causes of industry localization, where one main reason behind spatial clustering was that a local milieu with many separate firms within one line of business does offer a 'constant market for skills'. Given that we have largely failed to show empiri-cally that there is intense interaction going on between *firms* – whether traded or untraded – in regional clusters, could it be that spatial clusters of similar and related industries exist not because they make up a localized industrial system, but rather because they provide efficient labour markets for specialized skills?

Arguably we know too little of the dynamics of the local and regional labour markets of Hollywood, the City of London or Silicon Valley. Not to mention the hundreds and thousands of not so spectacular spatial clusters where we find substantial numbers of similar and related firms that normally exper-ience limited local interaction among themselves. Almeida and Kogut (1999) show that this research avenue may be rewarding. In a study of patent holders in the semiconductor industry, they find that Silicon Valley is clearly unique in terms of inter-firm mobility. The level of intra-regional mobility is found to be high, while the extent of inter-regional moves is much smaller. They suggest that innovative ability in this industry is strongly tied to the career paths of innovative individuals (see also Scott, 1988b; 1992). This is not to assume that firms in general regard labour mobility as a positive location factor. Indeed, most firms don't. Sometimes strong local informal institutions are established in order to limit local sharking of labour. However, the fact that most firms dislike losing employees to local competitors does not preclude that labour mobility is positive for the development of a spatial cluster in the longer run.

If there is something in this, we should (at least partly) shift focus from firms and industries to people and occupations and try to establish whether specialized local labour markets function differently from more diversified ones, and to analyse labour flows between firms (or e.g. from universities to firms) within and across spatial clusters. If it would be the case that such clusters are sites of more intense flows of people between firms, then we would perhaps have there the core explanation so far missing in the understanding of why spatial clustering occurs and why it should enhance the performance of firms.

Local and global circuits of interactive learning

Another argument, finally, departs from the simple notion that the local matters because this is where most people are bound to live most of their lives. Even in highly developed economies in the most globalized parts of the world, the group of people who have escaped the local and do indeed circle the world make up a tiny minority. Most people spend at least 250–300 days per year in the same place, following daily and weekly trajectories bringing them from home, to work, to a lunch restaurant, to a meeting downtown, or possibly in a neighbouring city, to some shopping area or child care institution and then back home. For sure, many people do regularly break out of this trajectory in order to go to conferences, business meetings or holidays in distant places, but most of the time they remain in the 'local circuit'.

Let us now assume that – apart from reading books and following media, which you could obviously do more-or-less regardless of location – people learn new things by interacting with others that they meet. Then we must assume that for most people, the overwhelming majority of interactive inputs come from the local circuit, while a tiny share comes from the global circuit. Some insight can emerge from applying the language of time-geography. How many and which people an individual will interact with during his or her daily, weekly, monthly or annual trajectory through time-space is, to use Hägerstrand's (1970) concep-tual framework, determined by three types of constraints. Capability constraints are essentially biological. Our daily activities are limited by the fact that we need to eat at regular intervals, sleep a number of hours and (in most cases) need 'a place to call home'. Coupling constraints define when, where and for how long individuals need to form 'timespace bundles' (i.e. to be within reach of one another in timespace, e.g. by being in the same place at the same time) in order to carry out whatever production, consumption or transaction they are involved in. Domain constraints, finally, refer to all kinds of institutional barriers that regulate access to and usage of certain timespaces.

Of course, the interactions taking place in any particular local milieu may be subject to constraints that differ in character and magnitude (based on such things as the size, density and physical and institutional structure of the place in question). Still it seems reasonable to assume that the potential for interaction between individuals based in the same local milieu (defined here, for the sake of convenience, as a local labour market) is different than for individuals located in different local milieus.

The simple proposal here is that patterns of interaction (and consequently interactive learning processes) in the local circuit are qualitatively different from those in the global circuit. Interactions in the local circuit would seem to allow for much more flexibility. The likelihood of regularly meeting, and gradually developing a relation with, another person is infinitely much greater if he or she is based in the same local milieu. Presumably, local interactions are character-ized not just by being unstructured and unplanned, but also relatively broad and diffuse, sometimes unwanted and often seemingly of little immediate use. In the course of our daily trajectory, we 'run into' many others, only some of whom we actually intended or wanted to meet.

The interactions in the global circuit are very different. They do not come 'by chance'. On the contrary, they are often the result of devoted and targeted identification of specialist people who are believed to possess some pieces of knowledge of importance. Often people do invest a lot (time, money and energy) in order to establish these relations. Once in place, a long-distance relation to a colleague, e.g. a customer, a supplier or even a competitor, can en-able all kinds of interaction and knowledge flows, but the building up of such relations are costly and therefore their numbers will be limited.

This is partly in line with the claim that relational proximity need not coincide with the type of spatial proximity that follows from permanent co-location. The concept of communities of practice has been developed to capture the idea that dense communities of people can develop around a certain trade or profession and such communities can develop relations of trust and shared cognitions that allow for the sharing of tacit knowledge, despite being globally dispersed (Amin, 2000; Gertler, 2002c; also in this volume). This line of thinking has also found its way into the field of urban research, where cities are now being situated in the context of 'distanciated economic flows and networks' (Amin and Thrift, 2002). At the same time, such accounts do, in relation to the argument proposed here, tend to downplay too much the restrictions on human mobility that do persist also in today's allegedly hyper-mobile era.

One could think of life in academia as an illustrative example, which presumably is not that different from the industrial activities that economic geographers normally study. Most successful academics live in both circuits. They have a local everyday life interacting with students, administrators, heads, deans, vice chancellors, and colleagues in the department (and possibly in neighbouring departments if they are lucky enough to be located in a dynamic clu..., sorry, campus) as well as with family, friends and neighbours. At the same time, they have a global professional network across which they regularly interact through travelling and electronic media.

Given our assumption that new knowledge is developed from and during interaction with others, it would be strange to assume that only one of these circuits matter. I would suggest that the relative importance of the local and global circuit is essentially an empirical issue, which economic geographers could address. There are some reasons why one could hypothesize that the local circuit is indeed the most decisive. One is, simply, that the local influences are so dominating in quantitative terms that they will make a real difference. Another has to do with the fact the relations between actors in the local milieu, generally, will be characterized by more of cognitive proximity and a higher degree of trust. A third argument is that, in certain respects, the global circuit is equally accessible to everyone regardless of location (given, of course, that there is access to information and communic-ations technologies and transportation infrastructure at a certain level) and could therefore not be a source of distinctive comparative advantages. On the other hand, most would agree with the proposition that the academic scholar with a well-developed worldwide network of specialized and highly qualified peers is likely to be more successful than the colleague who dwells exclusively in the local milieu – even if that milieu happens to be good. To take this argument yet another round, one could hypothesize that the quality of the local knowledge structure in turn is to some extent a function of the quality of the global connections that the individual actors in the local milieu have collectively managed to develop.

If there is something to this, we have here another possible answer to the question of the role of the local. If the local milieu, where people are bound to live their everyday trajectories, is resource rich (in this context, rich in knowledge, information and skills relevant for the line of business where they happen to work), it is more likely that they will develop the skills that will help them prosper. In a local milieu with many actors with related and complem-entary skills and competences, it is possible to develop a more detailed division of labour. Each actor can, thanks to the presence of the others, become more specialized. Such specialization will presumably not just raise product-ivity, but also enable and encourage the build up of specialized global relations.

IN SUMMARY

Three main propositions have been advanced in this chapter. The first is that research on clusters and clustering should acknowledge that functionally defined industrial systems, i.e. industry clusters, are rarely confined to or contained within narrowly defined local milieus. Therefore, we should not expect to find intense local interaction between firms and institutions even in cases when we face spatial clusters of similar and seemingly related activities. Secondly, it has been proposed that the 'cement' of spatial clusters lies in the fact that such milieus form the basis for well-functioning markets for competences and skills, i.e. specialized labour markets. Thirdly, when approaching spatial clusters from the point of view of understanding how such milieus become sites of learning and knowledge creation, we need both theoretical and empirical analyses of the different qualities of local and global interaction. In approaching that task, a fruitful point of departure is presumably to drop the assumption that 'the more local interaction between firms and institutions the better'. Rather we should acknowledge that most interactions at the individual level, for reasons that were identified decades ago by time-geographers, are indeed local. Therefore, the challenge, if we want to understand how a local milieu can develop and sustain international competitiveness, is to analyse how such milieus can develop links to and tap into sources of specialized knowledge, wherever in the world they happen to exist. Interactions in local milieus are fascinating and interesting, but understanding global connections is at least equally important.

PART THREE

REFIGURING GLOBAL RULES

Chapter 10

MAKING GLOBAL RULES
GLOBALIZATION OR NEOLIBERALIZATION?

Adam Tickell and Jamie Peck

INTRODUCTION:

globalization or neoliberalism?

Globalization and neoliberalism are both perplexingly ubiquitous phenomena. The orthodox understanding of globalization is based on a notion of increasingly borderless market extension, an apparently all-encompassing 'condition' in which market rules and competitive logics predominate, while the political leverage of nation-states recedes into insignificance. Meanwhile, the political project of neoliberalism represents a parallel attempt not only to visualize a free-market utopia, but to *realize* these self-same conditions, as the downsizing of nation-states enlarges the space for private accumulation, individual liberties and market forces. Perhaps not surprisingly, globalization and neoliberalism are often elided and entangled. Advocates of both tend to emphasize the need for corporations, governments and social actors to adjust to the new 'realities' of global competition; both envisage the role of markets in terms of apolitical, largely benign and integrating forces; both portray governmental bureaucracies and social collectivities as impediments to economic progress; and both actively anticipate world-wide processes of upwards convergence – a 'race to the top' – culminating in the establishment of a new orthodoxy or 'era'.

Globalization and neoliberalization are often elided, both historically and analytically: historically, because they are both held up to be creatures of the latter third of the twentieth century, and analytically, because both processes are typically ascribed a kind of ubiquitous causal agency by both celebrants and critics alike. Neoliberal politicians will often invoke globalization, as a signifier of powerful and in many respects unstoppable market forces, in order to advance the case for government sell-offs and privatization, fiscal austerity, financial and labour market deregulation, trade liberalization, welfare cutbacks and so forth. Simultaneously, critics of these policies and opponents of free-market globalization will often pointedly label all such phenomena as evidence of a creeping neoliberal (or, sometimes, American) hegemony. What the former are trying to depoliticize the latter seek to repoliticize – and the use of the label 'neoliberal' suits the latter because it is they who wish to underline the *political* origins and character of the programme.

The economic narrative of globalization and the political script of neoliberalism are both, in a sense, compellingly simple. They describe a new world order of untrammelled markets and competitive freedoms in clean lines and uncompromising terms. Implicitly or explicitly, they portray countervailing interests as unrealistic and outmoded. There are, of course, *always* alternatives to neoliberal political projects, just as there is a vast array of possibilities for organizing and regulating the global economy. Yet it is one thing to recognize the limitations and silences of the scripts of neoliberal globalism, quite another to move beyond these in a way that is conceptually sound and empirically informed. Economic geographers, in particular, have long opposed 'flat earth' conceptions of neoliberal globalization, based on unmediated market hegemony, cultural homogenization, institutional convergence and the associated assertion of a 'one best way' in corporate governance, economic regulation and social policy. As Peter Dicken's work has conspicuously demonstrated (see Yeung and Peck, in this volume), globalization tendencies are very much 'real' ones in legal, material and discursive terms, but they neither produce unitary outcomes nor do they erase local and national differences in business culture, corporate strategy or government policy. Globalization, in other words, produces its own geography, the resultant unevenness reflecting more than simply residues of 'pre-global' social formations, but an array of politically mediated forms of integration into a complex and changing global economic system. More than this, the 'outcomes' of globalization are politically negotiated and mediated; they are not predetermined by some 'hidden hand' of international market forces (see Gertler, Brenner and Hudson, in this volume).

A parallel set of arguments can be marshalled in the case of the global political project of neoliberalism. Despite having become the ideological 'commonsense' of the times (see Bourdieu and Wacquant, 2001) and contrary to its casual elision with 'Americanization', neoliberalism is far from a monolithic, undifferentiated project. It too has a geography, with its centres of discursive production (in places like Washington, DC, New York City and London), its ideological heartlands (like the US and the UK), its constantly shifting frontiers of extension and mediation (such as South Africa, Eastern Europe, Japan and Latin America) and its sites of active contestation and resistance (think of Seattle, Genoa, Cuba,...). But it would be a mistake to imagine that neo-liberalism is being projected out, in an unmediated form, from the Anglo-American 'heartland'. While the 'Washington consensus' clearly reflects the geo-economic interests of the global North, as a policy programme, it has been forged, adapted and reshaped in a wide variety of global contexts – from 'shock treatment' strategies in Chile, through to the Asian financial crises and its ramifications, to 'structural adjustment' programmes in developing countries. Neoliberalization, like globalization, should be thought of as a contingently realized *process*, not as an end-state or 'condition'.

Adequate conceptions of neoliberalism must be attentive both to its 'local' mediations and institutional variants and to the 'family resemblances' and causative connections that link these together. Just as there are no 'pure' markets, only markets shaped by different configurations of legal frameworks, social conventions, power relations, institutional forces and such like, so also there is no 'pure' form of neoliberalism, only a range of historically and geog-raphically specific manifestations of neoliberalization-as-process. Neoliberal politics, by the same token, are always hybrid politics – reflecting the balance of local political forces, sources of active resistance and institutional legacies, amongst other things – even though they will often appeal to ostensibly universal concepts like market efficiency and individual freedom. There is not one set of 'neoliberal states' and another set of 'non-neoliberal states'; neither are there straightforward transitions from one to the another, but a range of nat-ionally and locally specific and qualitatively differentiated forms of neoliberalization. So the process of neoliberalization has exhibited a *qualitatively* different form in, say, the Scandinavian welfare states and the Asian developmental states, in which neoliberalized elements have been grafted onto quite distinctive state structures. The outcome is not one of neoliberal homogenization, but a continuing process of uneven development within which neoliberal impulses are intensifying. Rather than seeking to measure in quantitative terms which nation-states are 'more' or 'less' neoliberal, the task is to trace the uneven effects of the neoliberalization process over time and space (see Brenner, in this volume).

Neoliberalization, in other words, refers to the process of political-economic change, not just the institutional outcome. It would be quite wrong, then, to anticipate a process of simple convergence towards a singular neoliberal state form, reproduced transnationally. Rather, neoliberalization – a.k.a. the extension of market rule and disciplines, principally by means of state power – defines the tendential form, content and trajectory of the restructuring process. For example, welfare restructuring processes in a range of countries and local contexts might be described as conforming to a common 'workfarist' pattern (in the sense that they are systemically orientated to deregistration, diversion and labour-market inclusion), even though they are being prosecuted in quite different institutional contexts and with markedly variable outcomes. It follows that if neoliberalization, defined in this way, represents the common tendential *form* of the restructuring process, then the outcomes of this process will be contingent and geographically specific, since they are working themselves out in a non-necessary fashion across an uneven institutional landscape.

Paralleling the body of work that has sought to deconstruct the project of globalization (see Amin, 1997; Dicken et al., 1997a), the overlapping and intersecting project of neoliberalization also needs to be subject to critical scrutiny (see Brenner and Theodore, 2002; Larner, 2000; Swyngedouw, 2000). This chapter represents a contribution to this latter strand of work, which is at a much earlier stage of development than the now 'mature' critiques of globalization. Complementing an earlier paper on the spatiality of neoliberalism (Peck and Tickell, 2002), our objective here is to develop a schematic account of the historical development of the neoliberal project, focusing explicitly on some of the leading geographical edges of its uneven development. This begins in the 1970s when the neoliberal credo was initially stitched together from diverse strands in free-market economics, individualistic philosophy and anti-Keynesian politics. These subsequently mutated into a series of state projects and restructuring programmes during the 1980s, most notably in the form of 'structural adjustment' initiatives in developing countries and various 'national neoliberalisms' in the US, New Zealand, the UK and elsewhere. Most recently, during the 1990s and beyond, there has been a period of consolidation and extension, marked by the ascendancy of the 'Washington consensus' as an hegemonic policy fix and the further morphing of the project itself into a range of socially ameliorative and authoritarian forms. Our goal here is not to present an 'end of history' account of the remorseless rise of neoliberalism to a position of global dominance, but instead to provide a sketch of its uneven ascendancy that draws attention both to its historically/geographically differentiated nature and to its complex evolution in response to internal and external pressures. In conclusion, we reflect on the political and theoretical integrity of the neoliberal project, within which deep contradictions and systemic vulnerabilities coexist with what, for many, is a perplexing degree of political adaptability, institutional durability and organizational creativity.

CHARACTERIZING NEOLIBERALIZATION

Neoliberalization is defined here, in process-based terms, as the *mobilization of state power in the contradictory extension and reproduction of market(-like) rule.* This is a far more complex and multifaceted process than the notion of 'deregulation' implies, for it has involved the development of new forms of statecraft – some concerned with extensions of the neoliberal market-building project itself (for example, trade policy and financial regulation; see Glasmeier and Conroy, in this volume), some concerned with managing the consequences and contradictions of marketization (for example, penal and social policy). It also implies that the boundaries of the state and the market are blurred and that they are constantly being renegotiated (see O'Neill, 1997). Neoliberalization is the dominant contemporary means through which such 'boundary adjustments' are being made and rationalized, with far-reaching consequences for both states and markets. The basic policy package through which the neoliberal project has been projected, often known as the Washington consensus, has been characterized by Standing (2002: 26) as follows:

1   trade liberalization

2   financial market liberalization

3   privatization of production

4   'deregulation'

5   foreign capital liberalization (eliminating barriers to foreign direct investment)

6   secure property rights

7   unified and competitive exchange rates

8   diminished public spending (fiscal discipline)

9   public expenditure switching (to health, schooling, and infrastructure)

10   tax reform (broadening the tax base, cutting marginal tax rates, less progressive tax)

11   a 'social safety net' (narrowly targeted, selective transfers for the needy)

12   flexible labour markets.

Neoliberal discourses tell a deceptively simple story about the logical, historical and philosophical superiority of markets, and of individualized and privatized economic relations more generally, coupling this with a concerted political programme to defend and extend the spaces of market rule. But this simplicity really is deceptive in that it is very often necessary for neoliberals to deploy state power and public authority in pursuit of these goals, underlining the reality that 'markets' are not naturally occurring phenomena or spontaneously actualizing systems. More often than not, they have to be made, steered and policed.

> It is axiomatic, according to neo-liberalism, that the absence of state intervention *is* the market, that market failures are never failures of the market *per se* and, therefore, they can only ever be failures of the state ... The political consequence of this view is the drive to deregulate... [Yet] the neo-liberalist vision of 'less state' is entirely illusory. Neo-liberalism is a self-contradicting theory of the state. The geographies of product, finance and labour markets that it seeks to construct require *qualitatively* different, not less, state action. Neo-liberalism is a political discourse which impels rather than reduces state action. (O'Neill, 1997: 291–2)

The reality of neoliberalism is therefore never as pure as its free-market rhetoric, while its oft-stated disdain for all things governmental sits uneasily with its actual practices of statecraft. While neoliberal projects are guided, in an often quite programmatic way, by a set of clearly articulated discursive precepts regarding the primacy of markets, individualism and the private sphere, their concrete manifestations are invariably more prosaic, contradictory and institutionally cluttered. More often than not, the *practice* of neoliberalism has little to do with laissez-faire deregulation – letting markets do their work, 'underseen' by an absentee state – but instead is associated with the extensive deconstruction and reconstruction of institutions, often in the name of or in the image of 'markets'. Thatcher-era privatizations in Great Britain, for example, did not result in the simple displacement of state-owned or government-controlled operations with freely functioning markets – say for energy, telecommunications or transport – but a tangled web of state-regulated oligopolies, profit-orientated enclaves and pseudo markets (see Feigenbaum et al., 1998). It follows that neoliberalism is anything but a coherent, singular and unchanging project. It has undergone a series of mutations and transformations during the course of the past quarter-century or so. On the one hand, this reflects a number of 'internal' movements in dominant philosophies, preferred tech-niques and policy priorities, expressed within the neoliberal project itself; on the other hand, shifts in the strategy and content of neoliberal policy have also occurred in response to encounters with 'external' obstacles and oppositional forces.

Even though neoliberals typically appeal to universal and ahistorical conceptions of market primacy, and while these continue to provide the basic philosophical parameters of neoliberal policy development, the concrete shape of the 'project' itself has always been institutionally variable, both across space and through time. In other words, the project has evolved, and it has evolved unevenly. The political 'success' of neoliberalism, it must be emphasized, was never guaranteed in advance, nor was its course clearly mapped out by the project's founding ideologues and vanguard politicians. Instead, the ascendancy of neoliberalism – from the ideological critiques of the 1970s, through the national-state projects of the 1980s, to the global hegemonic dominance of the 1990s and beyond – was in retrospect a faltering expansion through a number of qualitatively distinctive phases, each with its own emphases and weaknesses, each with its own geographical heartlands and frontiers. In the process, the 'project' has become ever more variegated and pervasive. It has spread beyond those national and local enclaves that were its basing points and control centres during the 1970s and 1980s (such as the US and the UK) to envelop much of the global South and a range of countries where social democracy had much deeper roots.

Yet neoliberalism did not diffuse in an unchecked and unmediated form – from national experiment to global fix. In the course of its uneven international expansion, it has mutated into a number of historically and geographically distinctive forms. The task of mapping the historical geography of neoliberalism remains in its infancy. As a contribution to this task, we sketch out here some of the ways in which the neoliberal project has evolved in the period since the 1970s. At the risk of over-simplification, we schematize the principal shifts that have taken place in terms of an uneven movement from 'proto-neoliberalism', referring to the inchoate origins of this ideological strategy in the period until the late 1970s, through the 'roll-back neoliberalism' of the deregulationist 1980s, when the primary focus was placed on dismantling social-collectivist and Keynesian-welfarist institutions, to the most recent phase of 'roll-out neoliberal-ism', during which active institution building has been increasingly in evidence in the context of an uneasy marriage of under-regulated markets and authoritarian governance. The three phases are summarized in Table 10.1.

Roots: proto-neoliberalism

It is salutary to reflect on the relative novelty of neoliberal thinking. Before the accession of the Thatcher government in the UK in 1979 and the Reagan government a year later, the Keynesian post-war consensus remained largely intact, albeit creaking, in most of the advanced industrial nations. This was root-ed in the widely accepted principles of full (male) employment, the maintenance

Table 10.1 Modes of neoliberalization

	Proto-neoliberalism	Roll-back neoliberalism	Roll-out neoliberalism
Periodization	Pre-1980	1980s/early 1990s	Since early 1990s–
Mode of political practice	Extra-state project	Statization	Hegemony/state-building
Dominant discourses	Anti-Keynesianism/state failure	Small government/ deregulation	Paternal state/free economy
Key institutions	Liberal think tanks	Governing parties	State cadres
Mode of political rationality	Ideological critique	Ideological project	Technocratic management
Sources of resistance	Keynesian orthodoxy	Organized labour	Cyber-activitists
Intellectual frontier	Monetarist economics	Supply-side economics	Bourgeois sociology
Totemic figures	Milton Friedman, Hayek, Pinochet	Thatcher and Reagan	Clinton, Blair, Schröder, Greenspan
Principal agents	Theorists, philosophers	Vanguard politicians, political appointees	Policy-functionaries, technopols
Intellectual elite relations	Confrontation	Conciliation	Cooptation
Service delivery	Spending cuts	Privatization	Marketization
Labour regulation	Crisis of full employment	Mass unemployment	Full employability
Fiscal posture	Stagflation	Tight money/liberal credit	Persistent deflation
State finances	Fiscal crisis	Systemic indebtedness	Debt repayment
Geographic heartlands	Chicago	London and Washington, DC	Brussels, London, and Washington, DC
Geographic frontiers	Santiago	Brussels	Paris, Berlin, Hong Kong, Singapore, Johannesburg
Spaces of resistance		British cities and coalfields, North American rustbelt	Anti-globalization confrontations, France, Malaysia
Scalar constitution		National	Glocal
Financial discipline	Inflation	Structural adjustment	Standards and codes
Ethic	Individualism	Amoral marketization	Moral authoritarianism

and extension of a welfare safety net, extensive industrial and labour-market intervention, social and spatial redistribution and demand-side macro-economic management. These state forms evolved under conditions of complex, mutual dependence with the dynamics of the Fordist economic expansion. In the context of an institutionalized social contract that exchanged workplace compliance for rising real incomes, this expansion was grounded in a series of virtuous circuits of output, productivity, consumption and profit growth. But during the 1970s, these virtuous growth dynamics began to unravel, triggered *inter alia* by the emergence of competition from Newly Industrializing Economies, a slowdown in productivity growth and profits in the Atlantic Fordist zone, the oil shocks, the internationalization of capital flows, rising inflation and unemployment and growing labour-union militancy. Both the political legitimacy and the financial viability of the Keynesian settlement were called into question as economic growth faltered, tax revenues fell, spending commitments spiralled and as a whole series of corporatist institutions publicly failed.

It is important to emphasize, however, that the developing political and economic crisis around the Keynesian-welfare state during the 1970s was not *destined* to translate into the ascendancy of neoliberalism. Instead, this was a period in which the fragments of a neoliberal 'state project' were being woven together. In the mid-1970s, neoliberalism was perceived as a kind of experimental shock treatment suitable perhaps for basket-case economies of the developing world, but hardly a basis on which to establish a viable governing ideology. Beneath the surface, however, lay an increasingly influential intellectual movement, which for some time had been pursuing a 'revival of the liberal creed' (Taylor, 1997) in the face of the extant Keynesian orthodoxy. Decidedly outside the intellectual and political mainstream during the three decades of the Fordist expansion after the Second World War, a loose network of proto-neoliberals had been earnestly engaged in a reconstruction of an alternative world view rooted in the centrality of market relations. With intellectual roots traceable back to Adam Smith and David Ricardo, the catalyst for this revivalist movement was Hayek's (1944) excoriating analysis of collectivism. Along with Milton Friedman, Hayek would go on to form the Mont Pélérin Society in 1947, which was dedicated to opposing the 'collectivist ideologies' of nationalism, socialism and fascism. At a time when the dominant political-economic discourse was resolutely statist and interventionist, this advanced guard for neoliberalism was laying the intellectual foundations for an alternative ideological project. The Mont Pélérin Society would subsequently spawn an international network of pro-market intellectuals with formative links to the most influential neoliberal think-tanks, such as the London-based Institute for Economic Affairs and the Washington-based Heritage Foundation and Cato Institute (Desai, 1994).

A parallel and related transformation was also underway within the discipline of economics. Milton Friedman and his acolytes at the University of Chicago had been working for some time to develop a distinctly *neo*classical form of economic theory, again in opposition to the dominant Keynesian rationality of the economics establishment (Dezalay and Garth, 2002). For two decades, this endeavour remained marginal to economics, but by the early 1970s, the tide began to turn.

The social turbulence and economic uncertainties of that period may have caused many in the social-science community to question the fundamentals of capitalism, but the ascendant mode of orthodox economic theory proposed an entirely different explanation: the economic difficulties of Western Europe and North America were consequences of *government* failure. Markets were not, according to this view, the cause of the problems of the 1970s, they were their cure. And the policy solution should come in the form of the rigorous imposition of market rule – first and foremost through monetary means. The intellectual architects of the neoclassical counter-revolution, peripheral figures in the economics profession since Keynes, now found themselves fêted. Hayek and Friedman both won the Nobel Prize for economics in the mid-1970s. And Friedman would become a very 'public' economist – as Keynes had before him – developing a significant presence in the financial press and on the TV networks, where the simple nostrums of market fundamentalism played well (Delazay and Garth, 2002).

Think-tanks represent a third strand in this emerging neoliberal network, explicitly focused as they have been on the construction of a new 'commonsense' around problem diagnosis and policy formation. In the strife-torn environment of 1970s Britain, leading economic commentators, policy advisers and financial journalists were being successively won over by neoliberal arguments, questioning the very rationales for government intervention and proposing new responses based on privatization and deregulation. The London-based Institute of Economic Affairs assumed a central role in these neoliberalizing policy debates (Desai, 1994). Much of the traffic in intellectual ideas during this period was trans-Atlantic, reflecting the deeper roots of individualistic, market-based philosophies in the US. Here, a newly animated network of think-tanks, most of which had substantial financial backing from the business community, were promoting the message that economic freedom was a *sine qua non* for a free society. These included the Heritage Foundation, founded in 1973 with an endowment from the Coors brewing dynasty; the Cato Institute, founded in 1977 with a highly libertarian agenda; and the American Enterprise Institute, which with a rather different remit had been founded in 1943, but in the mid-1970s converted to a strong version of conservative economics.

Yet for all the apparent intellectual coherence of the emergent neoliberal project, prior to the 1980s it remained largely confined to the realm of ideas rather than policies. In many respects, connections between neoliberal intellectuals and the murky world of mainstream politics were distant, if not hostile. The overthrow of Salvador Allende in Chile in 1973, however, presented a 'real world' test for the emergent neoliberal doctrine. The incoming Pinochet regime would seek to purge the vestiges of socialism from the Chilean political economy by embarking on a radical course of textbook neoliberalization. Central figures in the administration of this shock treatment were the 'Chicago Boys', a network of University of Chicago-trained Chilean economists who came to occupy key positions in think-tanks, financial institut-ions and government departments (Dezalay and Garth, 2002; Valdes, 1995). The neoliberal strategy for Chile involved the construction of a 'dualist state', based on the co-existence of extreme (and violent) political authoritarianism with rationalities of economic freedom, realized through the abolition of price controls, deliberate exposure to international competition, an extensive and rolling programme of privatization, and the elimination (sometimes literally) of the labour movement. Although Chilean authoritarian liberalization was indeed associated with a transformation in the economy, the underlying causes of this transformation remain hotly contested (Schurman, 1996; Valdes, 1995). Few dispute that it occurred at the expense of massively widened socio-economic inequalities.

Yet despite the uncompromising nature of its neoliberal experiment, Chile remained somewhat peripheral to the broader objective of dismantling the Keynesian commonsense in its strongholds – the advanced industrial nations of the global North. More portentous, in many respects, were events in Britain in 1976. Faced with an apparent crisis in the balance of payments, a collapse in the value of sterling and seemingly ineffective intervention in financial markets, the incumbent Labour Government was forced to apply 'cap in hand' (as commentaries of the time put it) to the International Monetary Fund (IMF) for emergency financial aid. Although later it became clear that the dire analysis of the British economy was distorted and exaggerated (Cairncross, 1994), the IMF's intervention was an important moment in the neoliberal ascendancy for at least three reasons. First, the very representation of the crisis established the kind of 'external' conditions appropriate for a neoliberal regime shift. As the argument was framed, in fact, this shift was *necessitated*, for to paraphrase Margaret Thatcher, there was no alternative. Second, the IMF's conditions amounted, in effect, to a first-world form of 'structural adjustment', of the kind subsequently imposed by the IMF and the World Bank on many developing countries during the 1980s – constrain inflationary growth, reduce the money supply, cut governmental spending and impose market-orientated rules. And third, the British crisis vividly focused attention on the supposed failures of Keynesian economics, providing a platform for the successful electoral camp-aign of Margaret Thatcher in 1979. Together with the election of Ronald Reagan in the US the following year, this marked a real turning point: in these two countries during the 1980s, the neoliberal project would take on a programmatic quality, as the fledgling ideology was melded with significant state power.

REACTION: ROLL-BACK NEOLIBERALISM

In the early 1980s, neoliberalism moved from being an emergent state project to a dominant state strategy. In London and Washington, the principal circuits of neoliberal influence spilled out beyond the think-tanks and seminar rooms and into the institutions of government. Although many governmental practices were slow to change, the rhetoric of political leadership under Reagan and Thatcher shifted dramatically. The talk was of individual freedoms and entrepreneurial flair, of government in the interests of ordinary citizens rather than big institutions, of low taxes and bureaucratic roll-backs. Amongst other things, the 1980s witnessed the wholesale privatization and retrenchment of public services; the progressive liberalization of credit; significant middle-class tax cuts; the extensive reworking of industrial relations law and practice; the removal of boundaries between different branches of the financial sector; the relaxation of regulations across a swathe of industrial sectors; and major confrontations with labour unions. While there were strong elements of pragmatism and incrementalism here, it is important to recognize that the reform process was guided by a clear, if still evolving, set of programmatic principles – minimize the size of government, make space for competitive forces, enlarge the scope and reach of the private sector, (re)distribute wealth on the basis of market principles, breaking down labour unions and other 'anti-market' or 'anti-competitive' institutions and so forth. These priorities, which allowed politicians of the right to seize the mantle of radicalism, were definitionally and diametrically opposed to those of the Keynesian era.

Labour markets and industrial relations were to be one of the first battle-grounds where the Reagan and Thatcher governments focused on the restoration of the 'right to manage'. This called for the establishment of a new industrial relations climate in which the rights and roles of labour unions would be dramatically curtailed. Both governments engineered confrontations with powerful labour unions in order to signal the change in policy and, more importantly, to inflict long-term damage on the labour movement. Key moments here came with the air traffic controllers' strike in the US and the steel and coal strikes in the UK. The legal and policy framework was also reorganized so as to be less accommodating to labour-union interests, to restrict strike actions and to limit the scope for collective bargaining. The image guiding these shifts was that of an individualistic and competitive labour market in which wages reflected human capital attributes rather than negotiating strength. Somewhat euphem-istically, these were described as 'flexible' labour markets in contrast to the 'rigid' and institutionalized employment systems of the Fordist period.

Alongside the flex-labour market offensive came changes in monetary policy. The Friedman 'monetarist' doctrine held that government profligacy led directly to inflation, proposing remedies in the form of reductions in govern-ment spending in particular and restrictions on the money supply more general-ly. Monetarist theory and its associated techniques of macro-economic management really constituted the cornerstone of neoliberal economic thinking during the early 1980s. But the cost of these policies was extremely high, both in terms of unemployment – famously dismissed as a 'price worth paying' to combat inflation by a British Treasury minister – and, perhaps more pertinently, in terms of reduced economic capacity. While single-minded monetarism fell out of favour amongst neoliberals by the mid-1980s, the obsession with inflation control most certainly did not. It is now embedded within the formal remit of the Bank of England and remains a central plank of the 'Wall Street consensus' in the US (Bluestone and Harrison, 2000). On both sides of the Atlantic, economic policy has become increasingly orientated to the interests of the financial markets, the roots of which are traceable back to the monetarist 'shock treatment' of the early 1980s.

But the neoliberal advance of the 1980s was not restricted to countries like the UK and the US, as totemic as politicians like Reagan and Thatcher proved to be.

More important for the project as a whole were developments at the international level, where one by one most of the leading international economic agencies fell under the influence of neoliberal ideologies. Waves of neo-classically trained economists, many of them with Chicago degrees, were recruited to supranational institutions such as the World Bank, the OECD and the IMF. Echoing the circumstances of Britain's crisis-induced bail-out in the 1970s, the nature of the development 'policy problem', and its remedies, began to change. During the 1970s, third-world countries had borrowed cheap, dollar-denominated funds from western banks awash with petrodollars in order to pursue westernized development agendas. The oil shocks of the 1970s and the monetarist-inspired ratcheting up of interest rates under Reagan and Thatcher in the early 1980s undermined this unsustainable situation. At the IMF, the new tenet stipulated that third-world countries' economic problems would only be addressed by the international community if their respective governments complied with 'structural adjustment programmes' (Rupert, 2000; Taylor, 1997). Loans would be advanced on the condition that recipient nations agreed to implement an unyielding neoliberal reform programme, typically including austerity measures, high interest rates, cuts in government spending in fields like education and health, the liberalization of foreign trade and investment and the privatization of public enterprises. While World Bank policies were somewhat more variable, they too relied heavily on neoliberal measures during the 1980s.

In its own terms, neoliberalism's roll-back phase was a phenomenal success. Markets and institutions were transformed as the politically legitimate remit of state intervention was redrawn. 'Deregulated' financial markets assu-med a new steering, policing and disciplining role, rewarding national states that aimed to cut social spending while managing the economy in a prudent fashion, and punishing those who ventured to think otherwise. Furthermore, whilst a few countries (particularly New Zealand, the UK and the US) publicly embraced neoliberal precepts, the state-assisted project of market integration across Europe, culminating in the single currency, acted to extend neoliberal policies to a new frontier.

For all this, by the early 1990s, the political reproducibility of neoliberal-ism in its heartlands looked far from secure. While the Reagan and Thatcher eras may have left very long legacies in terms of the broad parameters of policy and the structure of political discourse, their immediate end was less auspicious. In both the UK and the US, the debt-financed economic expansions of the 1980s ended in a predictable pattern of overheating and recession. The consequences included a shakeout of white-collar labour, corporate downsizing, a severe debt overhang and a cultural critique of greed and opportunism. The raw form of 1980s neoliberalism seemed to have accentuated the economic cycle, dumping the costs of economic adjustment onto those at the bottom, and even middle, of the income distribution, while exposing social and spatial inequalities to the point that they threatened to become politically combustible. Less charismatic politicians, albeit still of the political right, whose approaches proved to be more managerial than inspirational, were to succeed Reagan and Thatcher.

But if the moment of aggressive neoliberalism had passed in Britain and the US, this hardly signified the exhaustion of the project. Rather, the 1990s witnessed the effective 'normalization' of neoliberal modes of regulation, which increasingly came to constitute the taken-for-granted context for economic policy decisions. This shift to a more technocratic and managerial form of neoliberalism also reflected the changed circumstances in which the project was being advanced. 'New' policy problems had been created in the wake of the neoliberal offensive of the 1980s, including failures of deregulation and privatization on the one hand and large-scale under-employment and working poverty on the other, which now had to be managed even if they could not be 'solved'. And so the domestic political agenda in countries like the UK and the US began to change as the focus of attention shifted from establishing a neoliberal vanguard to managing the contradictions of earlier rounds of neoliberalization (Peck and Tickell, 2002; see also Brenner, in this volume).

PROACTION: ROLL-OUT NEOLIBERALISM

At the end of the 1980s, then, it may have appeared that the history of neoliberalism would prove to be nasty, brutish and short – no more than a destructive, (de)regulatory experiment. While this was actively anticipated in many quarters, the 1990s did not bring the collapse of the neoliberal experiment. On the contrary, the project acquired a *diffuse but consolidated* form, its central tenets having been absorbed into a truly hegemonic ideology in the sense that they now infuse 'mainstream' political discourses across much of the developed and developing capitalist world, while shaping the architectures of multilateral institutions and regional state structures (such as the European Union and the North American Free Trade Agreement). Emphatically, this does not imply either the end of history or the end of geography: the 'final triumph' of neoliberal ideologies has not been secured, while the reach and purchase of neoliberal methods of statecraft remain unevenly developed. Rather, what has happened is that the dominant *form* of neoliberalism has changed. If the 1980s witnessed a crude and shallow form of neoliberalism based on the roll-back of Keynesian-welfarist institutions and various experiments in market-making, the period since the early 1990s has witnessed a much deeper form of neoliberalization shored up by a new round of *neoliberal* state-building.

One of the more far-reaching effects of this deep process of neoliberal-ization has been the attempt to sequester key economic policy issues beyond the reach of explicit politicization. Economic decision-making has become too important to be left to democratically accountable politicians, evidenced by – for example – the granting of operational independence to the Bank of England or the extent to which the European Central Bank is protected from any effective scrutiny. Neoliberalism has also become more deeply embedded in international law, perhaps most vividly in the form of the World Trade Organization (see also Glasmeier and Conroy, in this volume). A neo-Ricardian conception of comparative advantage – which posits that every national-state must have a comparative advantage in *something* and that free trade is consequently an undeniable good – now permeates international economic relations. Under the WTO, long-standing bi-lateral agreements under which former colonizing powers in some way compensated their poor former colonies by giving preferential market access must disappear; other than in the most dire of medical emergencies, member nations must respect and pay for the rights of western pharmaceutical companies to patent drugs derived from third-world life forms; western agricultural combines control the rights of farmers to retain part of their harvest for replanting; and environmental considerations are typically shoved aside. As Rupert (2000: 49) has argued, one of the wider implications of such measures is that

> Even as people in locations around the globe are increasingly integrated into – and affected by – transnational social relations, neoliberalism seeks to remove these relations from the public sphere – where they might be subjected to the norms of democratic governance – and subject them to the power of capital as expressed through the discipline of the market.

At the same time as market logics have become naturalized, however, political *rhetorics* have changed. In response to the growing leverage of anti-globalization movements, some of the supranational institutions are tempering their discursive liberalism and increasingly stressing participatory politics, engaging with poverty agendas, or arguing that freer trade and market liberaliz-ation represent tools for the development of poor countries (for example, IMF, 2001), even while the powerful trading blocs maintain internal subsidies and market protection.

At the level of regulatory practice, there has also been a significant maturation of neoliberal policy networks. What Dezalay and Garth (2002: 30) call the 'dollarization' of after-Keynesian statecraft rests upon a complex of dense and intersecting networks – linking together US business schools and economics programmes, finance ministries and economic policy agencies, think-tanks and multinational agencies. Through these circulate a new breed of neoliberal *technopols*, 'strongly embedded in an international market of expertise modelled on the United States'. An extensive network of policy functionaries and technopols has progressively displaced the political appointees of the 1980s, reflecting important changes in the nature of the neoliberal project and the way in which it is prosecuted. Roll-out neoliberalism, in this sense, refers to both a technocratic turn in neoliberal practice and a qualitative shift in the dominant mode of neoliberal policy-making.

> In the early period as advisers to the government, the Chicago group devoted themselves to dismantling state regulations in accordance with the recipes that brought them to power. But, particularly after the excesses and scandals of the roaring 1980s, the priorities reversed. *In order to consolidate the policy gains of the neoliberal revolution, it appeared essential to reconstruct some level of regulation.* Far from being obstacles to the effectiveness of the market, therefore, the law and other supporting institutions now appeared as conditions necessary for its functioning. (Dezalay and Garth, 2002: 170, emphasis added)

Within the advanced industrial nations, too, neoliberal practices have undergone a form of transformation. Especially important, in this respect, is the new breed of 'social democratic neoliberals' (such as Blair, Clinton, Chretien and Schröder) who have demonstrated a commitment to ongoing marketization and public-private 'partnerships', in which the former absorbs the political-economic risks, while the latter absorbs the profits. While Third-Way rhetoric makes a great deal of its supposed break with neoliberalism, the more telling break is with the traditions of Keynesian welfarism. In fact, new forms of centrist politics are being formed around the foundations of neoliberal economics and its associated policy mix – flexible labour markets, free trade and minimalist government. A trademark of 'Third Way' political strategy, in fact, has been a focus on the 'downstream' consequences of economic liberalization – such as crime and social exclusion – coupled with a marked deference to the principles of 'sound financial management' and competitive globalization (for example, Giddens, 1998). Implicitly rejected is any serious engagement – intellectual or political – with the challenges of economic regulation and strategy. One of the quiet successes of neoliberalization has been to place these discussions practically 'off limits' in mainstream political discourse. Neoliberal policy positions establish a kind of taken-for-granted context for prudent political decision-making, the prerequisites for which increasingly include the maintenance of non-inflationary (usually slow) growth, a low taxation/low public-spending posture, resistance to deficit spending and government borrowing, and a commitment to increased competition in the delivery of public services, in the operation of labour markets and in the international trading system. Such normalized neoliberal conditions, of course, do not prevail in every single instance of state decision-making since political contingencies will sometimes override them (think, for example, of the pressure to halt the decline of public services in the UK, or the Bush administration's protectionist policy towards the steel industry), but it is a measure of the neoliberalized times that such sporadic policy initiatives are widely recognized to run against the grain of more fundamental commitments.

One of the essential characteristics of the more mature (or deeper) forms of neoliberalization that have evolved in the 1990s, then, has been the partial depoliticization of previously contested economic policy fields. Increasingly, these have been absorbed into the technocratic management routines of neoliberalized organizations (such as central banks or multilateral agencies) or they have become entrenched as effectively inviolable policy positions (for example, the desirability of flexible labour markets). A second key aspect of this phase of neoliberalization, however, relates to the new frontiers of 'active' policy-making, where purposeful forms of institutional roll-out are in evidence. Partly in response to the failures of previous forms of neoliberalization and partly in response to new policy challenges, neoliberals have found themselves increasingly engaged in proactive forms of statecraft and institution-building. Hardly a stable situation, what might be characterized as a new regulatory 'unsettlement' has been taking shape, with the effect of consolidating a series of neoliberal movements in political rationalities, policy conventions and modes of intervention. In substantive terms, this has included shifts from national macro-economic management to the facilitation of global economic integration; from the policy orientation of full employment to the new focus on full employability; from passive and redistributive welfare states to active and punitive 'workfare' regimes; from the governmental techniques of social-democratic intervention to those of Third-Way pragmatism; and from a predisposition to social and spatial redistribution to the acceptance (or even encouragement) of a Darwinian order of market distribution and naturalized inequality.

These conditions tend to be associated with Janus-faced state strategies – making, extending and managing markets in the broad field of economic policy, while responding to social and penal policy crises in an increasingly activist and authoritarian manner. The contemporary neoliberal state is a facilitative, market-managerial presence in matters of capital regulation, but adopts an ever-more aggressive, invasive and neopaternalist attitude towards the regulation of the poor. In this context, the political philosophy of roll-out neoliberalism can be neatly summarized in the following way:

> 'free,' that is, (neo)liberal and non-interventionist 'above,' in matters of taxation and employment; intrusive and intolerant 'below,' for everything to do with the public behaviors of members of the working class caught in a pincer movement by the generalisation of underemployment and precrious labor, on the one hand, and the retrenchment of social protect-ion schemes and the indigence of public services, on the other hand. (Wacquant, 1999: 338)

The roll-out of the penal state can be seen as a perversely logical response to the contradictions engendered by previous waves of deregulation and commod-ification, pointing to internal 'dynamics' of transformation within the process of neoliberalization. In fact, some of the same political narratives used to justify the roll-back of the welfare state are now being redeployed in order to legitimate new forms of regulatory roll-out, governance-making and proactive statecraft. Discourses of urban violence, dysfunctional welfarism, and ghetto pathology, for example, have been performing important ideological work in the processes of state restructuring and transformation, rationalizing the transition towards 'big stick' state strategies (Gilmore, 1998). It is in this context that the once-dominant social-welfarist orientation of Keynesian states is progressively giving way to, or been unevenly displaced by, new modes of governmental rationality based on penal management and punitive regulation, both of poverty and of poor subjects. Increasingly, these are the tasks that preoccupy those on the politicized front line of neoliberal policy-making, and not only in the global North. Chile, for example, recently announced the wholesale privatization of its unemployment insurance system, at the same time, in fact, as the Bush administration was seeking to intensify work requirements within the residualized welfare/workfare system in the US. These are the new frontiers of roll-out neoliberalism.

CONCLUSION:

global neoliberalism?

The casual elision of neoliberalism and globalism have resulted in conceptions of the former that are simultaneously inadequate and totalizing. While under-specified processes of globalization are often inappropriately ascribed a kind of omnipresent causal efficacy (Dicken et al., 1997a; also in this volume), so also it is necessary to be wary of holding neoliberalism responsible for all significant political-economic change over the past two decades. This would simply be a globalized version of the British left's generic complaint of the 1980s – 'I blame Thatcher'. Of course, there is a time and place for blaming Thatcher, or neoliberalism more generally, but no political or analytical purpose is served by doing this indiscriminately. Rather, there is a need to specify carefully the processes and mechanisms of neoliberalization, to understand its different institutional variants and to examine how these are interconnected through new, translocal channels of policy formation.

As an initial step in this direction, we have argued here for a qualitative understanding of neoliberalism in all of its spatially variegated, institutionally specific and historically changing forms. To this end, we presented a schematized sketch of the emergence of neoliberal forms of statecraft predominantly in the US and the UK. We have argued that neoliberalism has both a creative and a destructive moment and that any adequate treatment of the process of neoliberalization must explain how these moments are combined under different historical and geographical circumstances. In the stylized account presented here, it has been suggested that the destructive-deregulatory moment of neoliberalization was the dominant one in the 1980s and its creative-proactive moment has been ascendant in the subsequent period. If the early neoliberal state emerged, at least in part, as a response to the (real, perceived and constructed) failures of Keynesian welfare-statism, the late neoliberal state is increasingly consumed by the task of managing the contradictions of neoliber-alization itself – serial market and governance failure, social disintegration, environmental degradation and unsus-tainable growth.

And politics really matter here too. It is important to recognize that the neoliberalism did not spontaneously fill the vacuum created by the discredit-ation and failure of Fordist-Keynesian political-economic forms during the 1970s, though it certainly drew energy from these crisis conditions. There was nothing natural or inevitable about the ascendancy of neoliberalism; it is just as much of a political construction as the Keynesian order that preceded it. While our own reading of the early phase of deregulationist neoliberalism emphasized its inherent crisis-proneness, in retrospect the early 1990s was a tipping point for the neoliberal project rather than a terminal crisis. The roll-out phase that followed showed neoliberalism to be more adaptable and robust than was previously acknowledged. As we have defined it here, roll-out neoliberalism is an outcome of qualitative shifts in the manner in which state power is mobilized and new state expertises applied in pursuit of the wider objectives of extending and reproducing of market(-like) rule. In the process, neoliberal experiments have come to represent more than islands of exploratory statecraft; today they are deeply embedded in a complex of inter- and extra-local policy networks, institutional circuits and political structures. This implies a deep neoliberaliz-ation of spatial relations above and beyond the 'internal' neoliberal reform and reorganization of specific institutions, places or state structures (Peck and Tickell, 2002), a set of conditions that is stylized in Table 10.2. The 'in here' processes by which neoliberalization has progressively colonized different institutional spaces are therefore connected to its growing 'out there' presence as a coercive and constitutive force.

Table 10.2 Spaces of neoliberalization

		SPATIAL RELATIONS	
		Shallow neoliberalization … pursuit of locally specific strategies	Deep neoliberalization … inter-local logics and reflexive relations
I N S T I T U T I O N A L	**Roll-back neoliberalism** … its destructive and deregulatory moment	*Attacks on inherited Keynesian-welfarist structures, coupled with primitive deregulation of markets,* including: • monetarist macro-economic management • primitive marketization: dogmatic deregulation and privatization • place-specific assaults on institutional and spatial strongholds of welfare statism and social collectivism (e.g. social service cuts, deunionization)	*Extension of neoliberal strategies to the international domain,* including: • reductions in overseas aid • imposition of structural adjustment programmes • initial liberalization of financial markets and trading relations • external imposition of neoliberal strategies • intensification of coercive pressures emanating from international markets
F O R M S	**Roll-out neoliberalism** … its creative and proactive moment	*Proactive statecraft and institution-building in service of neoliberal goals,* including: • invasive moral reregulation of the urban poor • expansion of penal state apparatus and social control policies • continued crisis-management of deregulated and privatized sectors • extension of, and experimentation with, market-complementing forms of regulation • technocratic economic regulation within neoliberal parameters	*Normalization of neoliberal logics and premises in inter-local relations,* including: • 'fast policy' development through inter-local transfer, learning and emulation • appropriation of networking forms of governance and policy development • institutionalization of competitive globalism (e.g. WTO dispute resolution architecture, GATS proposals regarding public ownership) • posture of permanent adaptability and reflexivity in fields like urban governance and social/penal policy

Analytically as well as politically, there are dangers in overestimating the logic and coherence of the neoliberal project (see Larner, 2000), but there is a peraps more serious risk in underestimating its transformative, adaptive and crea-tive potential. The 'neoliberal state' is no longer, if it ever was, simply a deregulationist, absentee state, but has demonstrated a capacity to morph into a variety of institutional forms, to insinuate itself into, and graft itself onto, a range of different institutional settlements, and to absorb parallel and even cont-nding narratives of restructuring and intervention in response both to internal contradictions and external pressures. The project of neoliberalism is conse-uently plastic, evolving and changing. And the *process* of neoliberalization continues to produce institutionally, historically and geographically variable outcomes. The task of thoroughly mapping out the moving terrain of neoliberalization is only just beginning. The emergent geographies of neoliberalization – obfuscated and complex as they clearly are – represent some of the most significant global shifts of the contemporary era. Yet even if these conditions are being normalized politically, they must not be naturalized analytically. Neoliberalism was, and still is, politically constructed. Like that of Arturo Ui, its rise was, and is, resistible.

ACKNOWLEDGEMENT

We would like to thank Neil Brenner, Nik Theodore, Henry Yeung and participants at the WUN symposium at the University of Bristol in May 2002 for useful discussions around the issues in this chapter, responsibility for which remains ours.

Chapter 11

GLOBALIZATION:
FAUSTIAN BARGAIN, DEVELOPMENT SAVIOUR OR MORE OF THE SAME?
THE CASE OF THE DEVELOPING WORLD AND THE EMERGING
INTERNATIONAL TRADE REGIME

Amy Glasmeier and Michael Conroy

INTRODUCTION AND MOTIVATION

Economic geography should rest at the core of all discussions of globalization. Globalization, whether economic, political, social or cultural, is inherently the analysis of the spatial spread of interconnectedness and interaction. In spite of this natural link, few geographers are present in the high-profile global debates about key issues in the current wave of globalization. It has become, by default, a terrain dotted with aspatial analysts such as economists and lawyers. One important exception is Peter Dicken's work and writings that can be found cited amidst discussions of the causes and effects of globalization (see also Yeung and Peck, in this volume). Peter's writings and public engagements incorporate some of the most important problems facing the contemporary evolution of the global system, from the emergence of new mechanisms for global governance to increasingly frequent street protests against the anti-democratic nature of that governance. Indeed, citations to his work by academics, business professionals, government officials and citizen activists highlight his success in inserting an economic geographic view to an increasingly worldwide conversation about the causes and consequences of economic globalization.

In the spirit of Peter's contributions to the evolving discourse on globalization, we take up one issue embedded in much of his work, which is the causal connections between global processes and emerging regulatory regimes. Like discussions found in his benchmark text, *Global Shift*, we attempt to integ-rate two interactive discussions of the evolving global trading regime. We first offer an analysis of historical context preceding the current era of globalization, which emphasizes the immediate post-Second World War debates and discusses the institutions that resulted ostensibly to facilitate development and prosperity around the world. We argue that an overarching dilemma of that post-war era was the reluctance of the developed countries to alter significantly or abridge the highly unequal access to wealth creation that characterized the twentieth-century global economy. After the Second World War, the debate about the need for global development was evident even as the Bretton Woods institutions were created. Yet in their fundamental formulation they were born captive to developed countries' interests and never achieved the distributive goals set forth in the rhetoric of the early post-war years. In effect, developing countries of the world have been waiting more than fifty years – with increasing impatience – for the dividends presumed to accompany increasing trade liberalization. Current public discord around globalization should then come as no surprise.

We then discuss the emergence of the new trade regime following the passage of the Uruguay Round of the General Agreement on Tariffs and Trade (GATT) and formation of the World Trade Organization (WTO). This analysis interrogates the extent that the new trade context ameliorates or in some way overcomes some of the limitations noted in the history of GATT. This analysis brings into sharp relief the more controversial and problematic elements of the new found global system. It highlights the need for careful and embedded re-search that extends the reach of economic geography further into the intersec-tion of policy and process. By way of introduction, we feel it is important to state that we believe that globalization is real; constituting a process that reflects an increasing level of economic, political and cultural integration across time and space. In our view, the process is inherently spatial and very much tied to the increasing geographic integration of economic and political networks linking organizations, individuals and institutions. Like Held et al. (1999), we see the process of globalization as consisting of both enabling and constraining influen-ces that reflect the actions of individuals restrained by the effect of structures. We also hold that globalization presents both opportu-nities and challenges for members of communities to mobilize in response to globalization's harsher effects.

In the remainder of this chapter, we explicitly focus on the evolving trade regime and its circumscriptive effect on economic geography. We see the value of an economic geographic perspective in helping to disentangle fundamental challenges to community sovereignty and environmental integrity resulting from the implementation of a new trade regime. In our mind, it is not a question of whether there is a role for economic geographers in the emerging debates, but rather whether the discipline is willing to act affirmatively to help sort out the effects of increasing economic integration.

To make sense of what the emergence of the WTO and its legalistic framework of dispute settlement means for global trade negotiations, a quick review of trade policy in the post-war era seems essential. We recognize that a sweep of the twentieth century is dangerous in as much as it is possible to misplace unintentionally in time or entirely leave out critical events and therefore misdate important junctures. At the risk of such errors, this review will be simplified in the extreme, emphasizing key elements leading to the breakdown of the Bretton Woods agreement and the emergence of a public discourse about the need to let the 'rest of the world in' for meaningful discussions of global trade practices (see also Tickell and Peck, in this volume).

WAITING FOR GODOT

The early 1990s were watershed years for the publication of books and articles on the meaning and significance of the long-awaited passage of the Uruguay Round of the GATT. For non-GATT watchers, the most recent round's passage probably meant little more than a final resolution of an insufferably long process of arcane debate about minute differences in text interpretations of Byzantine trade negotiations. The difficulty of bringing the Uruguay Round to a conclusion says volumes about the post-war trade regime and its increasingly tattered form on the eve of the signing of the Marrakech Agreement in 1994. And even with its passage, only momentary relief prevailed. For the creation of the World Trade Organization, as prescribed at Marrakech, and the implement-ation of the Agreement's hundreds of clauses can only be said to be a rocky road that has so far been serviceable. Yet some would argue that it is dangerously close to collapse. There is some irony in this state of affairs because it was the near-collapse of GATT that led to pressure to enter into the Uruguay Round and establish the WTO (Shoch, 2001). Current circumstances can only be said to reflect the adage, 'Be careful what you wish, for you might actually get it'.

Wading through numerous perspectives on the post-1970s era of trade negotiations leads us to conclude that in the final analysis, passage of the Uruguay Round and the creation of the WTO were show-stopping necessities. Still being digested, the implications of GATT's passage continue to unfold. But to understand their significance requires stepping back in time to consider what the Bretton Woods institutions were anticipated to combat and what they were expected to promote starting some fifty years ago (see Aaronson, 2001 for a detailed chronicle of the role of community activism around trade-related policy in the US during the twentieth century).

Post-war trade policy: Bretton Woods

Reflecting on the creation of post-war economic institutions and the basis of their success, Eichengreen and Kenen (1996) point out four factors that were critical in promoting and maintaining post-war economic stability. First, during the crucial first years of the post-war era, the US was sufficiently well positioned economically to make the necessary side payments to assure adherence to the mission of economic stability and democratic reform among key countries worldwide. It made substantial investments in Asia, particularly in South Korea, Taiwan and Singapore, to ensure the countries would adhere to and promote democratic principles (Conroy and Glasmeier, 1994). These funds often came in the form of direct payments and military investments, but they also consisted of trade-related investments by US firms, protected by federal government insurance. It can be fairly said that the rise of the Asian Newly Industrializing Economies is explained, in part, by national policies associated with stabilizing a volatile region after the Second World War (Amsden, 2001; Conroy and Glasmeier, 1994; Glasmeier et al., 1993).

A second factor that helps explain the stability and success of the Bretton Woods institutions was the relatively small number and economic homogeneity of the signatories to the agreements. While nations of Latin America and Eastern Europe were in the background, the signatories to the agreements were the relatively few, rich and industrialized countries. Although some of the signatories had been devastated by the war, they had democratic institutions and variants on market economies that yielded a common set of interests.

Third, because of the economic isolationism of the interwar period, most signatory country economies were relatively closed. Trade had been choked off during the depression (Glasmeier, 2001; Gourevitch, 1990). Thus, given the closed nature of their economies, the signatory national governments could manage critical facets of their domestic economy and pursue policies that were supportive of domestic goals, such as full employment. Fourth, according to Eichengreen and Kenen (1996), there was an implicit agreement among citizens and the national governments of the Bretton Woods signatories that economic change would be managed and national policies would be created to maintain economic stability at home (Gourevitch, 1990).

While these four factors are only cryptically summarized here, the point of this discussion is to suggest that there were many stabilizing influences that made implementation of the Bretton Woods agreement possible. And yet, even as they held together, the unanimity brought about by these stabilizing influences was in fact vulnerable (Nye, 1996). As more countries experienced a process of industrialization, the US became less able to finance side payments to secure participation in global economic affairs. The will to see through not only the process of reconstruction of Europe but also a process of worldwide development was similarly eroded.

Bretton Woods institutions not up to the task of shared development

A closer look at the Bretton Woods institutions also points up inadequacies in their basic architectures. They were never intended to bring about worldwide development. In name only, the World Bank was not in fact set up to be a bank of worldwide development. Indeed as Eichengreen and Kenen (1996), as well as Nye (1996), persuasively suggest, the success of the Bretton Woods institutions was really confined primarily to the International Monetary Fund (IMF) and international monetary policy, and then for only a short time from the 1950s through to the end of the 1960s. To a lesser extent and for a shorter period, the same level of effectiveness can be said of the GATT.

The IMF had the clearest mission: to act as global gatekeeper responsible for overseeing the setting of exchange rates and the maintenance of balance of payments. The World Bank, in contrast, 'received comparatively little attention at Bretton Woods' (Eichengreen, 1998: 14). Both its capitalization and its opera-ting procedures were inadequate to meet the needs of post-war reconstruction. This inadequacy can be traced directly back to the reluctance of the US to pro-vide the necessary capital to fund the bank at a level where it could in fact act as a catalyst of development (Eichengreen, 1998; Eichengreen and Kenen, 1996). In saying this, such an admission in no way forgives the Bank for its exclus-ionary policies and 'one-size-fits-all practices'. There is even some suggestion that its culture bred a type of conservatism and a tendency to pursue self-preservation to the detriment of countries in need of assistance (Ul Haq, 1996). But still, the Bank could not 'reschedule a country's debt or manage intern-ational capital flows' (Eichengreen, 1998: 14). The developmental potential of the Bank was simply unrealized.

Thus, the historical record suggests, the world institutions emerging out of the Second World War were really limited in their scope and commitment to worldwide development. Moreover, a critical view of history would suggest there was no will to spread the benefits of market capitalism to the developing world. Indeed, the goal appears to have been to rebuild the tried-and-true market economies that had been damaged during the Second World War and little else. It should come as no surprise, therefore, that the Bretton Woods organizations were not up to the task of leading the world on a path of development.

The gradual collapse of the Bretton Woods agreement

The inadequacies of these institutions may also help to explain the growing sense of mounting frustration and the attendant disequilibrium that emerged in the early 1970s when the Bretton Woods compact began to break down. Developing countries wanted to be more than just repositories of natural resources. But the developed economies would have to give up markets in order to make room for developing economies' goods. This eroded the ability of the industrialized countries to ensure full employment. Thus a fundamental contradiction emerged. If developed economies promoted liberalization in turn they would have to become more open and therefore more affected by global economic trends.

The use of non-tariff barriers grew as markets were eroded. In turn, calls for a new economic order grew loud as developing countries saw their commodity-based economies decline as terms of trade continued to deteriorate (Baldwin, 1970; 2001). At the same time, as liberalization proceeded and cheap industrial goods penetrated developed country economies, the meaning and value of Bretton Woods institutions declined in the eyes of the ordinary citizen (Wolff, 1996).

The 1960s saw the end of unbridled enthusiasm (and expectations) for the smooth workings of the Bretton Woods institutions (Shoch, 2001). Balance of payments problems and dramatically different rates of economic growth among the industrialized countries began to erode the once united support for the Bretton Woods system. GATT was under attack by industrialized countries that experienced major increases in imports from Japan (Glasmeier et al., 1993; Shoch, 2001). These so-called side agreements challenged the fundamental authority and regulatory effect of the GATT. Pegged exchange rates became dysfunctional as countries faced with balance of payments problems resisted devaluation for fear of capital flight. The dollar came under increasing pressure and, by the early 1970s, the US could no longer honour the gold standard. The global financial system moved to floating exchange rates.

The 1970s was an era of tremendous change. The industrial economies devastated during the Second World War had by the late 1960s recovered and were producing competitive products for the world market. Key, less developed economies, especially those in Asia, were growing rapidly by successfully pursuing both import substitution and export promotion policies (Haggard, 1990).[1] Less developed economies, especially in Latin America, also practised import substitution development policies that yielded relatively high rates of growth for a period of time (Amsden, 2001). But the dramatic increase in exports during the late 1960s and the early 1970s raised new challenges to GATT, as developed country economies sought to keep low-cost imports out of their markets. In an effort to maintain full employment, industrial nations were increasingly resorting to non-tariff barriers to resist imports (Yoffie, 1993). Developing countries continued to register their complaints about the failure of previous trade rounds to reduce tariffs on imported agricultural products and unsuccessfully sought redress through various international mechanisms, including the United Nations Conference on Trade and Development (UNCTAD) (Ricupero, 2001).

Crisis brings economic adjustment in the absence of a plan

A great deal can be said of the ensuing 30 years when the fabric of Bretton Woods began to fray: first the abandonment of the gold standard, then the inflationary spiral set off by the oil crisis, followed by global recession and financial disequilibrium. The dark decade of the 1980s hit particularly hard those economies that had been relatively immune to the requirement of economic openness (Conroy, 1996). But the collapse of the global economy and the restrictions placed on borrowers by the IMF served to drive these economies into the ground. The effects of negative economic growth experienced by countries in Latin America and Africa are still very painful and rather immediate memories.

As for the Uruguay Round of the GATT, trade conditions had so deteriorated by the end of the 1980s that the passage of the Round, including the creation of the WTO, was considered by many to be a deal-breaking negotiation. If the Round failed to conclude, global trade was in jeopardy (Ricupero, 2001; Wolff, 1996). Countries had begun to form trade blocs. Some saw the movement behind the US-backed North American Free Trade Agreement as an alternative and potentially isolationist development in partial response to the EU integration (McConnell and McPherson, 1994; Saborio, 1992). The perception of trade in the US mattered a great deal. Free trade was no longer unequivocally accepted by US business interests operating domestically (Shoch, 2001). The 1970s and 1980s had polarized members of the business community and the larger society over the increasingly uncertain benefits of freer trade (Eichengreen and Kenen, 1996). Multinationals wanted it both ways: greater access and more protection. Citizen coalitions including farmers, unions and small businesses saw much to lose from pursuit of yet more open markets (Shoch, 2001). Other nation-states used import requirements and standards as a means of managing trade flows. Enforcement of the GATT had so deteriorated that exceptions and non-tariff barriers had increasingly become the rule (Baldwin, 2000). The US' use of unilateral measures to police imports, such as Super 301,[2] became a source of increasing frustration and retaliation by the nation's trade partners, who felt that the US wielded far too much power in the global trading game (Shoch, 2001).

The end of the Cold War reinforced, ironically, the growing separation among industrialized countries on issues of free trade. The collapse of the Soviet Union removed the geopolitical threat that had hung over the heads of the industrialized nations and created the historic unity found at the core of the belief system supporting the Bretton Woods institutions (Bergsten, 1996). Instead of greater unity, the collapse of the Socialist Bloc and the concomitant end to the Cold War fractured loose coalitions and projected the basis for new geopolitical arrangements based far more on geographic proximity than on the prospect of a distant military threat (Van Leenep, 1996).

The spectre of failure: GATT at a crossroads and the emergence of the World Trade Organization

After many false starts and more than eight years of negotiation, the Uruguay Round of the GATT passed in 1994 and was signed by more than one hundred nations in Marrakech, Morocco (Sampson, 2001). While much has been said about the Round's passage, references abound regarding the significant steps forward even in the face of some serious 'birth defects'.

For our purposes, the Round brought to life the WTO, an organization envisioned to facilitate trade negotiations, open up new avenues of trade and, importantly, to enforce agreements through a binding dispute settlement mechanism – a capacity heretofore lacking in the GATT. Also, with far less fanfare but perhaps with more far-reaching significance, the Round brought in as signatories the majority of the world's countries that had previously been bit-part players in the GATT. Earlier they had been excluded for all intents and purposes except when absolutely needed (Moore, 2001). Of equal weight but of different significance are areas covered by the charter, including services, intellectual property rights, subsidies and countervailing duties, standards, safe-guards, market access, developing country integration and dispute settlement that had previously been excluded from the GATT (Jackson, 1996; 2001).

Although recognized as having serious flaws, some only now emerging with clarity, nonetheless, the Round's passage created a new trading terrain hoped by many to solve some of the more onerous problems associated with the previous

GATT (Jackson, 1996; 2001; Sampson, 2001). And yet, in some critical and overarching ways, as we are only beginning to understand, the Round's passage sowed the seeds of popular dissent that now stands in the way of any hope of concluding a new Round, and may even threaten the very existence of the global trade regime as we know it (Annan, 2001; Arronson, 2001).

Was the creation of the World Trade Organization a Faustian bargain?

With so much dissension and conflict, how is it that the GATT passed and the WTO was created? Is current public outcry surprising? What happened to residual issues that failed to achieve much attention in previous rounds? To what extent is the current public displeasure with the WTO something new, or can we trace the concerns to an unremitting disregard for the fates of the marginalized nations that have made claims of the original post-war trade and development pact (Ul Haq 1996)?

There is no question that for the last 30 years, developing countries have voiced concern that issues of development, such as access to developed country markets for agricultural products and labour-intensive manufactured goods, have been ignored. In fact, marginalized countries signed on to the WTO precisely because they believed it was the best way to legitimize their claims about unfair trade. Participation was viewed as crucial in breaking the cycle of ineffectiveness felt by many developing countries (Moore, 2001; Ricupero, 2001; Sampson, 2001). In effect, their participation was bought with the belief that the long-standing and escalating problems of exclusion would be kept in check.

The WTO was assumed to offer a type of policing power needed to ensure that the benefits of trade finally did trickle down to the marginalized countries. For most developing countries, the Bretton Woods institutions offered little more than rhetoric and regulation. Thus one can only speculate about the array of reasons for the signing on by marginalized countries to the Uruguay Round of the GATT and the creation of the WTO. Perhaps the view was that things simply couldn't get any worse. Unfortunately, as the recent public demonst-rations suggest, they could and they have (Ricupero, 2001; Sampson, 2001).

Gary Sampson (2001), former director of the WTO Trade and Environment Unit, summed up the concerns of the marginalized countries, citizens worldwide and non-governmental organizations (NGOs) about the effects of increasing free trade and the implementation of the Uruguay Round of the GATT. First and foremost, the less developed countries feel that, as in the past, the WTO is an exclusionary organization. The WTO does not eliminate the unequal power in the global market. Indeed, given high entry and policing costs, developed countries can still throw their weight around and extract concessions from less developed countries simply because they have more resources to dedicate to the negotiations and larger markets to use as elements of leverage (Baldwin, 2001). Thus, in some important ways the WTO has not created a level playing field. As Kofi Annan (2001: 23), Secretary General of the UN, argues, less developed countries in many instances cut their tariffs more than developed countries, once again reinforcing unequal trade.

> Industrialized countries, it seems, are happy enough to export manufactured goods to each other, but from developing countries they still want only raw materials, not finished products. As a result, their average tariffs on the manufactured products they import from developing countries are now four times higher than the ones they impose on products that come mainly from other industrialized countries.

More importantly, in addition to concern that terms of engagement effectively differ little pre- and post-WTO, there is strong sentiment that new facets of the current trade regime in fact harm less developed countries. Particularly as they relate to such issues as intellectual property rights, bilateral requirements for accession, labour standards and rules of investment.

INCREASING GLOBALIZATION AND THE EMERGING TRADE REGIMES

In this section, we discuss questions that have arisen about the fundamental legitimacy of the global processes – and their governance – that affect local communities on two levels: first, in terms of the nature and structure of the representation they embody as these new processes and governance systems have been implemented; and second, in the troubling role being played by the office of the US Trade Representative (USTR) in the continued negotiation and implementation of the WTO rules. We explore the vulnerability of current conditions through the lens of the USTR and the TRIPs agreement on intellectual property rights. This analysis clearly shows that contrary to hopes and expectations of the LDCs (less developed countries), the emerging regime has sufficient distance still to travel before the inequality of power evident in the global trading system is eradicated. More than ever, economic geography is a vital optic to understand how newly emerging practices can and will alter the playing field of international engagement. Understanding the shifts in economic activity will undeniably require incorporating a sophisticated and nuanced appreciation of how trade policy is altering and in some critical ways structuring the meaning of access to markets to a greater degree than ever before. Global trade is by no means free trade and, if anything, regulatory oversight is increasing even as new participants and exceptions abound. As others have said before, it is at these times when uncertainty is high that the prospect for less versus more trade re-emerges as citizens and national governments increasingly question the legitimacy of the supposed gains from free trade. The spectre of isolationism always exists even as we push for more openness. As history demonstrates, the more unequal and uncertain the gains from trade, the higher the likelihood that alternatives emerge, be they trade blocs or more radical forms of closure. As the following case examples suggest, practices emanating from the most powerful players in the global economy provide little comfort to participants that have long waited to share in the gains from globalization.

Issues of fundamental legitimacy in global processes

A quick glance at popular press coverage would suggest there is the existence of a crisis of legitimacy surrounding processes of globalization. The vociferous opposition expressed by a broad spectrum of the public at major trade-related regional groupings are further reflections of disillusion, discontent and frustrated attempts to be heard. These protests are being organized by powerful national and international organizations that have found themselves excluded from the processes under which new rules of global governance are being written. They are challenging national government delegations as unrepresentative of the interests of many of their supposed constituents. Proposed compromises in global rules are being rejected as inappropriate by important portions of the public, whether they relate to liberalization of trade and investment or restriction of carbon emissions (Keohane and Nye, 2000; Esty, 2000; see also Tickell and Peck, in this volume).

Among delegations representing the 'developing countries'[3] in global processes, serious challenges have arisen about their representation, access to information and ability to participate on an equal plane with the wealthy countries. Demands from developing countries around 'implementation issues' threatened to block the launching of a new round of trade negotiations at the fourth WTO Ministerial Meeting in Qatar in November 2001 (*Bridges Weekly Trade News Digest*, 2001). These implementation issues are based ultimately on the failure of the developed countries to deliver the access to markets that developing countries *thought* they were being promised in the Marrakech Agreement that created the WTO. Many of these implementation issues focus explicitly on key questions of governance within the WTO and the evolution of dispute-settlement mechanisms, where many developing countries find them-selves increasingly disadvantaged in the emerging global system.

Would that legitimacy be enhanced if there were greater participation by local groups, local communities and other representatives of the disempowered? Frank Loy (2001), former US Undersecretary of State for Global Affairs, wrote recently:

> Much of the mistrust and many of the misconceptions about the WTO stem from its lack of transparency. When members of the public are excluded from a process, they tend to imagine the worst. If it is difficult to find out what the WTO is doing, then the organization is perceived as having something to hide.

Some have argued, in fact, that the WTO embodies an 'insidious shift in decision-making away from democratic, accountable forums – where citizens have a chance to fight for the public interest – to distant, secretive and unaccountable government systems, whose rules and operations are dominated by corporate interests' (Wallach and Sforza, 1999: 2). One doesn't need to agree with that proposition to perceive that the legitimacy of the organization is under serious question. Would a greater voice for local communities and the poor enhance that legitimacy?

The first response to this question is often indirect, an invocation of the 'impracticality' of encouraging a wider array of voices. Negotiation is already difficult enough, some say. Neither the WTO nor the secretariats of the major multilateral environmental agreements has the authority to give votes to anyone other than official national delegations. The locus of legitimacy may then focus more appropriately on the levels of participation in the development of each nation's negotiating and voting positions in the global institutions. It is at that level of national participation where local voices may be most needed.

Issues of legitimacy related to the WTO and global trade

Two case studies – the US Trade Representative's Office, and the Trade-related Aspects of Intellectual Property Rights known as the 'TRIPs' agreement, exemplify and highlight the challenges facing the once immutable global trading system. Actions on the part of the US in its role as a powerful actor in the global trade negotiations threaten to undermine world trust in the WTO. The practices of the USTR office and the practical effects of the TRIPs agreement are glaring examples of misuse and extensions of power that have led to suspicion and mistrust on the parts of nations of the global South as they consider the value and viability of negotiations around international trade.

The US Trade Representative's Office

It is difficult to meet with NGO representatives, developing country delegates or even WTO officials without encountering vociferous criticism of the manner in which the Office of the US Trade Representative manages its membership in and around the WTO. Although all are willing to concede that the USTR has a right to negotiate forcefully and to protect and enhance the interests of its citizens, both individual and corporate, the manner in which this business has been conducted is seen by many as threatening the fundamental legitimacy of the WTO.[4] First among the complaints are the attitudes of the US delegation in Geneva toward developing country issues:

1   The USTR frequently conducts bilateral negotiations with countries that seek admission to the WTO and, as a condition for US support for admission, it insists upon bilateral concessions for US exports and investments in those countries that go far beyond the WTO requirements for membership.

2   The USTR uses US market power in ways that undercut the fundamental principles of the WTO. It insists, for example, that all the developing countries open their textile and apparel markets immediately, at the same time it takes a 'non-negotiable' position that textiles and apparel are a 'sensitive' sector that cannot be opened in the US because of domestic politics.

3   The USTR exercises bilateral pressure on developing countries to deter them from exercising flexibility provisions built into the WTO, such as the compulsory licensing and parallel imports provisions for drugs which the WTO allows, but which US pharmaceutical firms oppose.

4   The USTR develops its negotiating positions in direct collaboration with key US industry representatives, resisting the participation of represent-atives of broader civil society including consumer organizations, environ-mental groups or representatives of other impacted constituencies.[5]

5   By using its influence with the UN to block proposed UNDP support for the South Centre, the USTR has sought to undermine the most important small research and training centre in Geneva that provides some of the very little capacity building available there for developing country delegations.

6   The USTR has refused to release documents that indicate the positions it has taken in most of its negotiations; it has now been sued for the release of its position papers sent to nations throughout the hemisphere in the negotiation of the Free Trade Area of the Americas.[6]

More generally, it appears that the USTR is considered obstructionist in the WTO Budget Committee, blocking, for example, additional WTO funding for capacity building among developing countries. It is seen as a 'bully' in processes for selecting a Director General, as a 'meddler' in the internal affairs of the WTO by insisting on having the right to select a Deputy Director General, and as hypocritical in pushing proposals for 'transparency' that it knows in advance will be rejected by most of the other delegations and that it does not practise in its own internal workings.

These actions by the USTR have significant impacts upon the legitimacy of the WTO, for they preserve imbalances of power within the organization advan-tageous to some US interests. They stifle alternative voices at almost every level. Citizens of the US might ask whether this is the nature of the representation that they seek to present in the global trade arena and whether greater transparency at home about the actions of their trade representatives might not increase the ability of the WTO to perform its functions fairly and effectively. They might also ask whether strengthening of local voices in the US might not contribute to the creation of an approach to trade positions by the USTR that reflect national preferences more adequately.

Extending claims over the use of natural resources: the trade-related intellectual property rights (TRIPs)

Recent actions to gain control over aspects of genetic knowledge associated with living materials in the natural world have raised serious questions regarding the WTO's ability to protect the intellectual property of nations of the global South from the predatory behaviour of transnational corporations. In its simplest and most direct form, the WTO agreements on TRIPs are little more than a generalization of 'patent rights', offering innovators an opportunity to obtain standardized global rights to their innovations.

TRIPs has become controversial for several reasons. First, it effectively extends patent law far beyond the levels of protection that the nationally negotiated laws of many of the member states would take it. Second, it is being used extremely aggressively by a small number of (largely Northern) trans-national corporations to capture rights to intellectual property that have been in the public domain for centuries and, in some cases, millennia (Shand, 2001).[7] For example, Article 27.3(b) of the TRIPs agreement covers patenting of 'life forms' (plants, animals and biological processes). As a result, a US firm has 'patented' basmati rice, an Indian staple for more than 400 years; and it seeks to recover royalties from all farmers in India and elsewhere who plant it, even when the seeds have been saved from previous harvests. Another US firm has purchased a few pounds of an unusually coloured yellow bean long grown in Oaxaca, Mexico; and it has obtained a patent on it, claiming to have 'stabilized' the production of that colour in the bean. It has also sued to collect royalties from US importers of the bean (Pratt, 2001). These are some of the more egregious examples of practices that are increasingly called 'biopiracy', which is seen by the Rural Advancement Foundation International (RAFI) as 'predatory on the rights of indigenous peoples and farming communities' (Raffinews, 2000: XXX).

According to James Orbinski (2001: 167), president of the International Council of the Nobel Laureate Doctors Without Borders, the application of the TRIPs agreement by pharmaceutical companies to restrict the availability of life-saving drugs worldwide constitutes 'trade practices that mean inequity and ultimately unnecessary suffering and death'. The TRIPs agreement includes provisions for 'compulsory licensing', which allow governments to seek generic production of patented drugs (including the payment of a reasonable licensing fee), and 'parallel imports', where governments can import drugs from the country in which they are available at lowest cost when firms discriminate from country to country by price. Some governments, led by the US government until recently, have brought bilateral economic pressure to bear on countries that propose to invoke those WTO provisions.

Developed countries are insisting that developing nations implement TRIPs-compliant changes in domestic legislation and practice within very short periods after joining the WTO. World Bank analyses suggest that the direct cost of this compliance, along with others required by the WTO, may exceed the average total amount of development assistance that developing countries are receiving (Finger and Schuler, 1999). For this reason, among others, delays in the implementation of TRIPs rules are one of the strongest demands of developing countries as a precondition for their support for launching further rounds of trade negotiations.

CONCLUSION: FAUSTIAN BARGAIN, NEW FOUND SAVIOUR OR MORE OF THE SAME?

Looking back historically, we can only conclude that the road ahead for less developed countries, countries of the global South, will only deviate from past practice to the extent that the new trade regime builds in room for the voice of citizens of the world. At present, as the previous section suggests, we are at a turning point – a juncture reminiscent but not exactly like the past. One direction will lead to a world of more equitable integration. Back in 1949, negotiations at Bretton Woods knowingly concluded with a club of nine countries seeking to rebuild and bolster a belief system based on an ideology of openness and participation, naively constructed retrospectively of a select few. It is improbable that such a closed group can operate in ironic isolation as before. The spectre of 11 September forecloses this option. But, should this be the ultimate outcome, such a path will result in a world of contention and dashed hopes. Of course, with any moment in history, there always is a middle ground, a little more fairness in return for a little bit less of other positions. If local communities worldwide are to create for themselves a realistic opportunity to benefit from globalization, they must first improve their understanding of the processes that engulf them. If they are to gain influence over those processes, they must build mechanisms for exercising their influence at both national and international levels. If they are to take better advantage of the opportunities that globalization may bring, they will need to establish learning frameworks that focus on successful experiences in comparable contexts. And if they are to gain greater ability to adapt to global processes, they will need support for that adaptation that is comprehensive, equitable and not demeaning.[8]

The meeting in Qatar at the end of 2001 brought into sharp relief the stakes governing the next round of trade negotiations. Subsequent meetings of the IMF, World Bank, and the G7 have in fact had high on their agenda global development: those key words implied, assumed, but never realized back in 1949. Everyone in the international trade community agrees that there will have to be concessions in order to forestall the prospect of an overly long and unfulfilling round of negotiation. Whether the threat of stalling will prove sufficient to ensure the rights of citizens of the global South are given true legitimacy is anyone's guess. No doubt parties to the discussion preparing for the arduous process will find themselves turning to the work of Peter Dicken to discern what's in it for them in a growing world of economic integration. No doubt the next edition of *Global Shift* has the prospect of being a reference text as negotiators attempt to fathom just what globalization will mean in the early decades of twenty-first century.

ACKNOWLEDGEMENT

This paper was originally prepared for presentation at the Clark University Conference on Globalization, supported by the Lier Fund on Studies of Globalization, Worchester, MA, October 2001. The contents of this chapter are ours and should not be construed as representing the views of the Ford Foundation.

NOTES

1   Import substitutions policies, popular after the Second World War, were programmes whereby nations protected infant industries through high tariffs and other barriers to entry; export promotion policies are instances where nations build capacity in goods where they have an export advantage.

2   Super 301 was an action undertaken by the American government in retaliation for governments thought to be dumping traded goods on the US market.

3   The term 'developing countries' seemed to disappear from international lexicons in the 1970s and 1980s; it was replaced by 'less developed' or 'less industrialized' or 'Third World', for the term 'developing' carried the connotation that they were simply countries that were 'getting wealthy' when the reality, as discussed below, is that many of them have been facing deepening imiseration for years. The WTO, however, reintroduced the phrase and we will follow that use for the sake of clarity.

4   Documentation of these perspectives is difficult, for neither WTO officials nor developing country delegates are willing to be cited directly for obvious reasons.

5   It took a successful 1999 lawsuit, for example, to force the USTR to accept a single environmentalist on the advisory panels for wood and paper products (see Griffith, 2000).

6   Earth Justice Legal Defense Fund, Press Release, 7 March 2001.

7   In some ways, the global race to capture intellectual property may be seen as a 'third wave' of colonization. If the first wave consisted of the colonization of territory from the fifteenth to the nineteenth centuries; the second wave may be seen as the conquest of financial markets by a small number of global corporations in the twentieth century. The conquest of global intellectual property, in the biotechnology field, for example, has been reduced to five or six global companies that have raced around the globe in less than ten years, competing to capture marketable rights (cf. Shand, 2001).

8   Peter Rosset of Food First, personal communication, 26 February 2001.

Chapter 12

'GLOCALIZATION' AS A STATE SPATIAL STRATEGY:
URBAN ENTREPRENEURIALISM AND THE NEW POLITICS OF UNEVEN DEVELOPMENT IN WESTERN EUROPE

Neil Brenner

INTRODUCTION:

urban entrepreneurialism through the lens of spatialized state theory

Since the late 1970s, the political geographies of urban governance have been transformed throughout western Europe. The welfarist orientation of urban political institutions that prevailed during the post-war boom has been superseded by a 'new urban politics' focused on the issues of local economic development and local economic competitiveness. This reorientation of urban governance has been famously described by Harvey (1989a) as a shift from urban 'managerialism' towards urban 'entrepreneurialism' and has been documented extensively (Hall and Hubbard, 1998). While the politics of urban growth have long been a central preoccupation within scholarship on US cities, the investigation of urban entrepreneurialism in western Europe has been intertwined with more recent debates on globalization, European integration and the crisis of the Keynesian welfare national state (Harding, 1997). In the face of geo-economic shifts such as the globalization of capital, the consolidation of the Single European Market and the decline of Fordist manufacturing industries, many western European cities have been confronted with intensifying socio-economic problems such as capital flight, mass unemployment and infrastructural decay. At the same time, as Keynesian welfare systems have been retrenched under pressure from neoliberal and 'Third

Way' national govern-ments, local states have been confronted with a more hostile fiscal environment in which they have been constrained to engage proactively in diverse economic development projects. Taken together, these transformations have underpinned an increasing neoliberalization of urban politics throughout western Europe, as the priorities of economic growth, territorial competitiveness, labour market flexibility, lean administration and market discipline have become increasingly naturalized as the unquestioned parameters for local policy experimentation (Peck and Tickell, 1994; Tickell and Peck, in this volume).

In recent years, analyses of the geographies of urban entrepreneurialism in western Europe have proliferated, filling a growing number of pages within international urban studies journals. Building upon these research forays, this chapter develops a state-theoretical interpretation of the uneven transition towards urban entrepreneurialism in western Europe. I argue that entrepreneur-ial cities represent key regulatory arenas in which new 'glocalized' geographies of national state power are being consolidated. Faced with the intensified globaliza-tion/Europeanization of economic activities and the increasing depend-ence of major capitalist firms upon localized agglomeration economies (see also Hudson, in this volume), these emergent glocalizing state institutions have mobilized diverse political strategies to enhance place-specific socio-economic assets within their territories. In contrast to the Keynesian welfare national states of the post-war era, which attempted to equalize the distribution of population, industry and infrastructure across the national territory, the hallmark of glocalizing states is the project of *reconcentrating* the capacities for economic development within strategic subnational sites such as cities, city-regions and industrial districts, which are in turn to be positioned strategically within global and European economic flows. This emergent strategy of urban reconcentration is arguably a key element within contemporary post-Keynesian competition states (Cerny, 1995) and has generated qualitatively new forms of uneven spatial development throughout western Europe. Crucially, however, the concept of glocalizing states is deployed here to refer not to a stabilized, fully consolidated state form, but rather to demarcate an important *tendency* of state spatial restructuring in contemporary western Europe. The process of glocalizat-ion will thus be theorized here as an emergent and deeply contradictory *state strategy* (Jessop, 1990) that hinges upon the spatial reorganiz-ation of state regulatory arrangements at multiple spatial scales.

In this chapter, I shall not attempt to document the transition to urban entrepreneurialism in western Europe or, for that matter, to differentiate among the diverse (national and local) political forms and institutional pathways through which this reorganization of urban governance has unfolded (Brenner, 2001). Instead, my primary goal is interpretive: I aim to outline a theoretical concept-ualization of *state spatial strategies* that illuminates the proliferation of local economic initiatives throughout the western European city-system during the last three decades. Like other contributions to this volume, this chapter emphasizes the uneven, politically mediated character of contemporary geo-economic transform-ations. The process of globalization is viewed here as a medium and expression of political strategies intended to undermine the nationally organized regulatory constraints upon capital accumulation that had been established during the post-war period. While such strategies have assumed diverse political-institutional forms around the world, they have frequently been oriented towards a rescaling of inherited national regulatory arrangements, leading in turn to an intensification of uneven development and territorial inequality at all spatial scales (Peck and Tickell, 1994). This analysis suggests that state institutions are playing a key role in forging the uneven geographies of political-economic life under early twenty-first century capital-ism. Thus conceived, states do not merely 'react' to supposedly external geo-economic forces, but actively produce and continually reshape the very institutional terrain within which the spatial dynamics of globalized capital accumulation unfold.

The next section elaborates a theoretical approach to the geographies of statehood under modern capitalism through a spatialization of Jessop's (1990) strategic-relational approach. I shall then outline an interpretation of the entrepreneurialization of urban governance and the 'glocalization' of state space in contemporary western Europe.

ON THE SPATIAL SELECTIVITY OF CAPITALIST STATES:
THEORETICAL FOUNDATIONS

While traditional accounts of statehood presupposed numerous geographical assumptions (Agnew, 1994), contemporary geo-economic and geopolitical transformations have generated an unprecedented interest in the geographical dimensions of state power (Brenner et al., 2003). As this burgeoning literature has emphasized, contemporary transformations have entailed a reterritorializ-ation and rescaling of inherited, nationally organized formations of state spatiality rather than an erosion of the state form as such. Much of this research can be situated within a broader body of social-scientific work concerned to counter mainstream globalization narratives by examining the ongoing reorganization of state apparatuses in the context of globalizing/neoliberalizing trends. Thus, among the many arguments that have been advanced regarding the institutional architectures of post-Keynesian, post-Fordist, workfare or competi-tion states, recent discussions of state spatial restructuring are characterized by a distinctive emphasis upon the new scales, boundaries and territorial contours of state regulation that are currently crystallizing. Insofar as the apparently ossified fixity of established formations of national state territoriality has suddenly been thrust into historical motion, contemporary scholars are confronted with the daunting but exciting task of developing new categories and methods through which to map the rescaled, reterritorialized and rebordered terrains of statecraft that have subsequently emerged around the world.

Surprisingly, much recent work on the production of new state spaces has proceeded without an explicit theoretical foundation. In many contributions to this literature, the geographical dimensions of state power are treated in descriptive terms, as merely one among many aspects of statehood that are undergoing systemic changes. Just as frequently, the causal forces underlying processes of state spatial restructuring are not explicitly specified. Consequ-ently, there is an urgent need for more systematic reflection on the specific political-institutional mechanisms through which states' territorial and/or scalar configurations are transformed from the stabilized *settings* in which political regulation unfolds into the *objects* and *stakes* of socio-political contestation.

These issues can be confronted, I believe, through an inquiry into the state's contradictory strategic role in the regulation of capitalism's uneven geographical development at various scales. To this end, I shall draw upon Jessop's (1990) strategic-relational approach to the state in order to interpret contemporary processes of state spatial restructuring as expressions of *spatially selective political strategies*. On this basis, I shall then examine the state-led political strategies that have underpinned the transition to urban entrepren-eurialism in post-1970s western Europe.

According to Jessop (1990), the capitalist state must be viewed as an institutionally specific form of social relations. Just as the capital relation is constituted through value (in the sphere of production) and the commodity, price and money (in the sphere of circulation), so too, Jessop (1990: 206) maintains, is the state form constituted through its 'particularization' or institutional separation from the circuit of capital. However, in his view, neither the value form nor the state form necessarily engender functionally unified, operationally cohesive or organizationally coherent institutional arrangements.

The value form is under-determined insofar as its substance – the socially necessarily labour time embodied in commodities – is contingent upon (1) class struggles in the sphere of production; (2) extra-economic class struggles; and (3) intercapitalist competition (Jessop, 1990: 197–8). According to him, therefore, the relatively inchoate, contradictory matrix of social relations associated with the value form can only be translated into a system of reproducible institutional arrangements through *accumulation strategies*. In Jessop's (1990: 198) terms, an accumulation strategy emerges when a model of economic growth is linked to a framework of institutions and state policies that are capable of reproducing it (see also Jessop et al., 1988: 158).[1] He proposes a formally analogous argument regarding the state form whose functional unity and organizational coherence are likewise said to be deeply problematic. To him, the existence of the state as a distinctive form of social relations does not automatically translate into a coherent, coordinated or reproducible framework of concrete state activities and interventions. On the contrary, the state form is seen as an under-determined condensation of continual strategic interactions regarding the nature of state intervention, political representation and ideological hegemony within capitalist society. For Jessop, therefore, the functional unity and organizational coherence of the state are never pregiven, but must be viewed as emergent, contested and unstable outcomes of social struggles. Indeed, it is only through the mobilization of historically specific *state projects* that attempt to integrate state activities around a set of coherent political-economic agendas that the image of the state as a unified organizational entity ('state effects') can be projected into civil society (Jessop, 1990: 9, 346). State projects are thus formally analogous to accumulation strategies insofar as both represent strategic initiatives to institutionalize and reproduce the contradictory social forms of modern capitalism.

On this basis, Jessop introduces the key concept of strategic selectivity, the goal of which is to develop a framework for analysing the role of *political strategies* in forging the state's institutional structures and forms of socio-economic intervention. Jessop concurs with Claus Offe's well-known hypothesis that the state is endowed with selectivity – that is, with a tendency to privilege particular social forces, interests and actors over others. For Jessop, however, this selectivity is best understood as an object and outcome of ongoing struggles rather than as a structurally preinscribed feature of the state system. Accordingly, Jessop (1990: 260) proposes that the state operates as 'the site, generator and the product of strategies':

1   The state is the *site* of strategies insofar as 'a given state form, a given form of regime, will be more accessible to some forces than others according to the strategies they adopt to gain state power' (Jessop, 1990: 260).

2   The state is the *generator* of strategies because it may play an essential role in enabling societal forces to mobilize particular accumulation strategies and/or hegemonic projects.

3   The state is the *product* of strategies because its own organizational structures and modes of socio-economic intervention are inherited from earlier political strategies (Jessop, 1990: 261).

In this manner, Jessop underscores the relational character of state strategic selectivity. The state's tendency to privilege certain class factions and social forces over others results from the evolving relationship between inherited state structures and emergent strategies to harness state institutions towards particular socio-economic projects.

The state strategies in question may be oriented towards a range of distinct socio-institutional targets. In particular, strategies oriented towards the state's own institutional structure may be distinguished from those strategies oriented towards the circuit of capital and/or in the mobilization of societal hegemony. In Jessop's terminology, the former represent *state projects* whereas the latter represent *state strategies*. State projects aim to provide state institutions with some measure of functional unity, operational coordination and organizational coherence. When successful, state projects generate 'state effects' which endow the state apparatus with an image of unity, functional coherence and organiz-ational integration (Jessop, 1990: 6–9). By contrast, state strategies represent initiatives to mobilize state institutions towards particular forms of socio-economic intervention (Jessop, 1990: 260–1). When successful, state strategies result in the mobilization of coherent accumulation strategies and/or hegemonic projects (Jessop, 1990: 196–219). While state strategies generally presuppose the existence of a relatively coherent state project, there is no guarantee that state projects will effectively translate into viable state strategies (Table 12.1).

In sum, rather than viewing selectivity as a pregiven structural feature of the state, Jessop insists that it results from a dialectic of strategic interaction and socio-political contestation within and beyond state institutions. In this view, ongoing social struggles mould (1) the state's evolving institutional structure and (2) the state's changing modes of socio-economic intervention, accumula-tion strategies and hegemonic projects. Just as crucially, the institutional ensemble in which this dialectic unfolds is viewed as the result of earlier rounds of political struggle regarding the forms and functions of state power. Accordingly, 'the state as such has no power – it is merely an institutional ensemble; it has only a set of institutional capacities and liabilities which mediate that power; the power of the state is the power of the forces acting in and through the state' (Jessop, 1990: 270). The conception of the state as a political strategy is thus intended to illuminate the interplay between these evolving institutional capacities/liabilities and the ensemble of social forces acting in and through state institutions.

Table 12.1 State projects and state strategies

STATE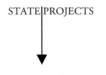PROJECTS	Initiatives to endow state institutions with organizational coherence, functional coordination and operational unity: they target the state itself as a distinct institutional ensemble within the broader field of social forces. • Target: state institutions • Possible outcome: '*state effects*'
STATE STRATEGIES	Initiatives to mobilize state institutions in order to promote particular forms of socioeconomic intervention: they focus upon the articulation of the state to non-state institutions and attempt to instrumentalize the state to regulate the circuit of capital and/or the balance of forces within civil society. • Target: the circuit of capital and/or civil society • Possible outcomes: accumulation strategies and/or hegemonic projects

Source: Based on Jessop (1990)

In an important extension of Jessop's framework, Jones (1997) has proposed that capitalist states are endowed with distinctive *spatial selectivities* as well. For Jones (1997: 851), spatial selectivity refers to the processes of 'spatial privileging and articulation' through which state institutions and policies are differentiated across territorial space to focus upon particular geographical areas. Building upon Jones' arguments, I would suggest that Jessop's strategic-relational approach can be fruitfully mobilized as the foundation for a spatialized conceptualization of state restructuring. The methodological lynch-pin of this conceptualization is the proposition that state spatiality is never a fixed, pregiven entity but, like all other aspects of the state form, represents an emergent, strategically selective and socially contested *process*. Just as radical approaches to urbanization under capitalism have long emphasized the proces-sual character of urban spatiality (Harvey, 1989b), so too is a dynamic, process-based understanding of state spatiality required in order to decipher the historical geographies of state restructuring under capitalism (Lefebvre, 1978).

Jessop's strategic-relational approach to the state provides a useful basis on which to develop such an analysis. As indicated, Jessop maintains that the organizational coherence, operational cohesion and functional unity of the state are never pregiven, but can be established only through political strategies. This argument can be fruitfully applied to the geographies of state power as well. From this perspective, the territorial coherence and interscalar coordination of state institutions and policies are never pregiven, but can be established only through political strategies to influence the form, structure and internal diff-erentiation of state space. Concomitantly, extant geographies of state institutions and policies must be viewed as the products of earlier strategies to reshape state spatial arrangements. The spatiality of state power can therefore be viewed at once as a site, generator and product of political strategies (MacLeod and Goodwin, 1999). State spatiality is forged through a dialectical relationship between (1) inherited patternings of state spatial organization and (2) emergent strategies to modify or transform entrenched political geographies. Building upon Jessop's strategic-relational theorization of the state form, state projects and state strategies, three equally fundamental dimensions of state spatiality under capitalism can be distinguished – the state spatial form, state spatial projects and state spatial strategies (Table 12.2).

1   *The spatial form of the state.* Just as the state form is defined by the separation of a political sphere out of the circuit of capital, the state's spatial form is defined with reference to the principle of territoriality. Since the consolidation of the Westphalian geopolitical system in the seventeenth century, states have been organized as formally equivalent, nonoverlapping and territorially self-enclosed units of political authority. Throughout the history of state development in the modern world system, the geography of statehood has been defined by this territorialization of politics within a global interstate system (Ruggie, 1996). Even in the current era, as national state borders have become increasingly permeable to supranational flows, territoriality arguably remains the most essential attribute of the state's spatial form, the underlying geographical matrix within which state regulatory activities are articulated.

2   *State spatial projects.* As indicated, the organizational coherence and functional unity of the state form are never structurally pregiven, but can be secured only through state projects that attempt to 'impart a specific strat-egic direction to the individual or collective activities of [the state's] different branches' (Jessop, 1990: 268). A formally analogous argument can be made with regard to the state's spatial form. Whereas territoriality represents the underlying geographical terrain in which state action occurs, its coherence as a framework of political regulation is never structurally pregiven, but can be secured only through specific state spatial projects that differentiate state activities among different levels of territorial admin-istration and coordinate state policies among diverse geographical locations and scales within (and, in some cases,

beyond) national borders. State spatial projects thus represent initiatives to differentiate state territoriality into a partitioned, functionally coordinated and organizationally coherent regulatory geography. On the most basic level, state spatial projects are embodied in the state's internal scalar differentiation among distinct tiers of administration. This scalar differentiation of the state occurs in conjunction with projects to coordinate administrative practices, fiscal relations, pol-itical representation, service provision and regulatory activities among and within each level of state power.

Table 12.2 A strategic-relational approach to state spatiality

STATE FORM
• The state is institutionally separated or 'particularized' from the circuit of capital

STATE SPATIAL FORM
• The state is organized as a territorially centralized and self-enclosed unit of political authority within an interstate system defined by formally equivalent political-territorial units

STATE PROJECTS
• The organizational coherence, functional coordination and operational coherence of the state system is never pregiven but is the product of particular programs and initiatives that directly or indirectly impact state institutional structures
• State projects represent attempts to integrate the ensemble of state activities around a common organizational framework and shared political agendas
• Target: state institutions
• Possible outcome: 'state effects'

STATE SPATIAL PROJECTS
• The geographical cohesion of state space is never pregiven but is the product of specific programs and initiatives that directly or indirectly impact state spatial structures and the geographies of state policy
• State spatial projects emerge as attempts to differentiate and/or integrate state institutions and policy regimes across geographical scales and among different locations within the state's territory
• Target: spatially differentiated state structures
• Possible outcome: consolidation of spatial and scalar divisions of regulation; uneven development of regulation

STATE STRATEGIES
• The state's capacity to promote particular forms of economic development and to maintain legitimation is never pregiven but is the product of particular programs and initiatives
• State strategies emerge as attempts to impose particular forms of socioeconomic intervention
• Target: circuit of capital and civil society• Possible outcome: accumulation strategies and/or hegemonic projects
... historically specific forms of STRATEGIC SELECTIVITY

STATE SPATIAL STRATEGIES
• The state's capacity to influence the geographies of accumulation and political struggle is never pregiven but is the product of particular programs and initiatives
• State spatial strategies emerge as attempts to mold the geographies of industrial development, infrastructure investment and political struggle into a 'spatial fix' or 'structured coherence' (Harvey, 1989b)
• Target: the geographies of accumulation and regulation within a state's territory
• Possible outcome: spatially selective accumulation strategies and/or hegemonic projects
... historically specific forms of SPATIAL SELECTIVITY

State spatial projects may also entail programmes to modify the geographical structure of inter-governmental arrangements (for instance, by altering administrative boundaries) or to reconfigure their rules of operation (for inst-ance, through centralization or decentralization measures) and thus to recalibrate the geographies of state intervention.

3   *State spatial strategies.* As we saw above, the state's capacity to promote particular forms of economic intervention and to maintain societal legiti-mation is never structurally pregiven, but can emerge only through the successful mobilization of state strategies. While the existence of a state project does not necessarily translate into the mobilization of a coherent state strategy, the consolidation of state strategies is a key precondition for the elaboration of accumulation strategies and hegemonic projects. Analogous arguments can be made to characterize the state's strategies to influence the geographies of industrial development, infrastructure invest-ment and political struggle. Just as states play a central role in the elaboration of accumulation strategies and hegemonic projects, so too do they intervene extensively in the geographies of capital accumulation and political struggle. In particular, states are instrumental in managing flows of money, commodities, capital and labour across national boundaries, in maintaining the productive force of capitalist territorial organization, in regulating uneven development and in maintaining place-, territory- and scale-specific relays of political legitimation. The resultant state spatial strategies are articulated through diverse policy instruments, including industrial policies, economic development initiatives, infrastructure investments, spatial planning programmes, labour market policies, regional policies, urban policies and housing policies, among many others. How-ever, the state's capacity to engage in these forms of spatial intervention and thus to establish a 'structured coherence' or 'spatial fix' for capitalist growth (Harvey, 1989b) is never pre-given, but can emerge only through the successful mobilization of state spatial strategies. The capacity to mobilize state spatial strategies does not flow automatically from the existence of state spatial projects. Nonetheless, it is only through the elaboration of spatial strategies that the state can attempt to influence the geographies of capital accumulation and political life within its jurisdic-tion. State spatial strategies are embodied in the territorial differentiation of specific policy regimes within state boundaries and in the differential place-, territory- and scale-specific effects of those policies. Whereas some state projects may explicitly promote this uneven development of regulation, this may also occur as an unintended side-effect of state action (Jones, 1997).

In short, Jessop's strategic-relational conceptualization of the state may be expanded into a 'strategic-relational-spatial' framework. In this conception, the geographies of the state under modern capitalism represent expressions of a dialectical interplay between inherited partitionings of political space and emergent state spatial projects/state spatial strategies that aim to reshape the latter. State spatiality can thus be conceived as a contested political-institutional terrain on which diverse social forces attempt to influence the geographies of state activity. Such struggles focus both upon the state's own territorial/scalar configuration (through the mobilization of state spatial projects) and upon the geographies of state intervention into socio-economic life (through the mobilization of state spatial strategies). In the remainder of this chapter, I shall mobilize this theoretical framework in order to analyse the role of entrepreneurial urban policy in the 'glocalization' of state space in post-1970s western Europe.[2]

FROM 'ENTREPRENEURIALIZED' URBAN SPACES TO 'GLOCALIZED' STATE SPACES

A preliminary characterization of glocalizing states was provided at the outset of this chapter. In contrast to the Keynesian welfare national state, with its project of equalizing the distribution of industry, population and infrastructure across national territories, glocalizing states strive to differentiate national political-economic space through a reconcentration of economic capacities into strategic urban and regional growth centres. The term 'glocal' – a blending of the global and the local – seems an appropriate label for these tendencies insofar as they involve diverse political strategies to position selected subnational spaces (localities, cities, regions, industrial districts) within supranational (European or global) circuits of economic activity. Although their political and institutional contours vary, strategies of glocalization have been mobilized by national states throughout western Europe (Swyngedouw, 1997). In each case, national economic space is being transformed into a 'glocalized composite' (Martin and Sunley, 1997: 282) as states manoeuvre to position their major urban and regional economies strategically within global and European circuits of capital. Table 12.3 summarizes contemporary glocalization strategies through an ideal-typical contrast to the strategies of spatial Keynesianism that prevailed under post-war capitalism.

In the present context, I shall build upon the approach to state theory developed above in order to interpret the 'entrepreneurialization' of urban policy during the post-1970s period as a key medium and expression of glocalization strategies. In this conceptualization, the glocalization of state space has unfolded through an uncoordinated constellation of political strategies – composed, in turn, of contextually specific state spatial projects and state spatial strategies – that have significantly recalibrated the relations between national and subnational scales of state regulation. Entrepreneurial urban policies have arguably played an essential role in animating this process of state rescaling.

Historical-geographical context

Glocalization strategies must be understood in relation to the dominant state projects and state strategies that immediately preceded them. Spatial Keynesianism was the dominant framework of state spatial regulation during the Fordist-Keynesian epoch throughout western Europe (Martin and Sunley, 1997). Its overarching goal was to redistribute resources to lagging or peripheral regions and thus to promote balanced urbanization throughout the national economy.

Table 12.3 Two strategies of state spatial regulation: spatial Keynesianism and glocalization

	STRATEGIES OF SPATIAL KEYNESIANISM	STRATEGIES OF 'GLOCALIZATION'
Geo-economic and geopolitical context	•Differentiation of global economic activity among distinct national economic systems under 'embedded liberalism' and the Bretton Woods monetary system• The polarization of the world system into two geopolitical blocs under the Cold War	• New global-local tensions: global economic integration proceeds in tandem with an increasing dependence of large corporations upon local and regional agglomeration economies • The end of the Cold War and the globalization of US-dominated neoliberalism
Privileged spatial target(s)	• National economy	• Major urban and regional economies
Major goals	• *Deconcentration* of population, industry and infrastructure investment from major urban centres into suburban and 'underdeveloped' rural peripheries • *Replication* of standardized economic assets and investments across the national territory • Establishment of a nationally *standardized* system of infrastructural facilities throughout the national economy • Alleviation of uneven development within national economies: uneven spatial development is seen as a *limit* or *barrier* to industrial growth	• *Reconcentration* of population, industry and infrastructure investment into strategic urban and regional economies • *Differentiation* of national economic space into specialized urban and regional economies • Promotion of *customized*, place-specific forms of infrastructural investment oriented towards global and European economic flows • Intensification of interspatial competition within and beyond national borders: uneven spatial development is seen as a viable *basis* for industrial growth
Spatio-temporality of economic development	• 'National developmentalism': development of the entire national economy as an integrated, autocentric, self-enclosed territorial unit moving along a linear developmental trajectory	• 'Glocal developmentalism': fragmentation of national economic space into distinct urban and regional economies with their own place-specific locational assets, competitive advantages and developmental trajectories
Dominant policy mechanisms	• Locational subsidies to firms • Local welfare policies and collective consumption investments	• Deregulation and welfare state retrenchment • Decentralization of social and economic policies and fiscal responsibilities

		• National urban policies and spatially selective investments in advanced infrastructures • Place-specific regional industrial policies and local economic initiatives
	• Redistributive regional policies • National spatial planning and public infrastructural investments	
Dominant slogans	• 'National development'; 'Balanced growth'; 'Balanced urbanization'	• 'Globalization'; 'Cities in competition'; 'Endogenous development'

1   As a state spatial project, spatial Keynesianism entailed the mobilization of intergovernmental policies to integrate local political institutions within national systems of territorial administration and public service delivery.

2   As a state spatial strategy, spatial Keynesianism entailed the mobilization of compensatory regional policies to extend infrastructure investment and industrial development into non-industrialized locations across the national territory.

This framework of state spatiality was destabilized during the 1970s in conjunction with processes of global economic restructuring, the crisis of the Fordist regime of accumulation and the retrenchment of the Keynesian national welfare state. In this context, traditional relays of national welfarism, regional redistribution and urban managerialism were increasingly seen as being incompatible with the need to reduce administrative costs, to enhance labour market flexibility and to promote territorial competitiveness in an increasingly volatile geo-economic system. Consequently, as of the late 1970s, glocalization strategies began to emerge, initially in the form of neocorporatist regulatory experiments intended to promote endogenous growth within declining industrial regions. In subsequent decades, glocalization strategies proliferated more widely as neoliberal, centrist and neocorporatist approaches to local and regional economic development were diffused across western Europe (Eisenschitz and Gough, 1993).

In contrast to spatial Keynesianism, which targeted the national economy as an integrated geographical unit, glocalization strategies promote the re-concentration of industrial growth and infrastructure investment within strategic urban and regional economies. Insofar as entrepreneurial approaches to urban governance represent one of the major regulatory experiments through which this goal has been pursued, they must be viewed as an essential component of glocalization strategies. As conceived here, therefore, glocalization strategies do not represent a unilinear resurgence of local economic governance, but entail, rather, a recalibration of national geographies of state power in ways that target the local and regional scales as strategic sites for regulatory experimentation.

A MULTIPLICITY OF POLITICAL AND INSTITUTIONAL FORMS

The common denominator of glocalization strategies is their privileging of subnational scales of state regulation and their promotion of local and regional economies as the motors of economic development. It should be emphasized, however, that the social bases, institutional forms and policy instruments associated with glocalization strategies vary considerably. In particular, the form in which glocalization strategies are articulated has been conditioned by inherited state structures (unitary vs. federal), inherited economic arrangements (the form of post-war growth), by national and/or regional political regimes (neoliberal, centrist or social-democratic) and by nationally specific pathways of post-Fordist industrial restructuring. A systematic comparative investigation of glocalization strategies in western Europe would therefore need to explore the diverse political-institutional forms in which they have been mobilized during the last three decades, even in the midst of their otherwise analogous spatial selectivities.

Glocalizing spatial projects and glocalizing spatial strategies

Glocalization strategies combine state spatial projects and state spatial strategies in distinctive ways.

1   As a state spatial project, glocalization has entailed initiatives to recon-figure the geographies of state institutions in ways that transfer new roles and responsibilities to subnational administrative levels – whether by recalibrating national and local institutional hierarchies, by introducing new scalar divisions of state regulation, by intensifying inter-administrative competition for state resources, by reconfiguring the administrative boundaries of subnational territorial units or by establishing entirely new subnational institutional forms.

2   As a state spatial strategy, glocalization has been associated with a variety of state-led regulatory experiments intended to resolve the crisis of the Fordist accumulation regime. Faced with an intensifying uneven develop-ment of socio-economic conditions within national economies, these state spatial strategies have attempted to enhance locally and regionally specific economic assets and to reconcentrate industrial development and infrastru-ctural investment within strategic cities, city-regions and industrial distr-icts. The national economy is thus to be fragmented among local and regional economies with their own place-specific assets and developmental trajectories.

In each case, the nationally organized economic and regulatory geographies associated with spatial Keynesianism are being superseded by political strat-egies intended to bolster the structural importance of local and regional scales of political-economic life.

An unstable, uncoordinated political strategy

Glocalization strategies are unstable, relatively uncoordinated and experimental. For, as with all forms of state spatial regulation, the geographical unity and interscalar coherence of glocalization strategies are never pregiven, but can exist only as outcomes of ongoing socio-political struggles to rescale state institutions, to endow state regulatory activities with particular forms of spatial selectivity and to promote particular accumulation strategies at determinate scales, locations and spaces.

1   The state spatial projects associated with glocalization generally lack internal unity and interscalar coherence. Following the destabilization of Fordist-Keynesian state space during the late 1970s, local and regional states throughout western Europe began to mobilize place-specific strategies of institutional restructuring in order to grapple with intensifying local social problems and enhanced fiscal austerity. The resultant state spatial projects more frequently represented a centrally induced fragment-ation of earlier frameworks of state spatial organization than a coordinated programME for restoring the state's geographical unity or for integrating policy initiatives across spatial scales.

2   The relationship between glocalizing state spatial projects and glocalizing state spatial strategies is deeply problematic. Even when glocalizing spatial projects have resulted in a significant recalibration of state spatial organiza-tion, the state's capacity to rework the geographies of capital accumulation is never guaranteed, but is an object of ongoing, strategically and spatially selective socio-political struggles.

Glocalization strategies and the 'creative destruction' of state space

The mobilization of glocalization strategies can be viewed as a double movement of socio-spatial transformation. On the one hand, glocalization strategies have entailed the partial destruction of earlier geographies of state regulatory activity, as projects of national spatial redistribution are increasingly abandoned or marginalized. On the other hand, glocalization strategies have also entailed the tendential creation of a rescaled scaffolding of state institutions and policies, as new frameworks for local and regional economic development are established. Crucially, however, this creative destruction of state spatiality must not be conceived as a complete replacement of one geography of state regulation by another. Instead, the forging of new geographies of state regulation occurs through a conflictual interplay between older and newer layers of state spatial activity, leading in turn to unintended, unpredictable and often dysfunctional consequences (Peck, 1998). Thus conceived, the diffusion of glocalization strategies in western European states has not simply 'erased' earlier geographies of state regulation, but has generated contextually specific, path-dependent rearticulations of inherited and emergent state regulatory practices at a range of geographical scales (Brenner, 2001). The glocalized formations of state spatiality that have crystallized during the last three decades represent an aggregate expression of this dynamic intermeshing of different rounds of state regulatory activity.

The uneven development of regulation and the regulation of uneven development

A new mosaic of uneven spatial development has crystallized in close conjunction with these glocalization strategies.

1   In contrast to the Fordist-Keynesian project of establishing a nationally standardized hierarchy of political institutions, the state spatial projects of the post-1970s period have entailed an increasing geographical differenti-ation of state regulatory infrastructures, systems of public service delivery and policy initiatives across the national territory. The uneven development of state regulation which results from these customized, place-specific regulatory strategies is an essential characteristic of glocalized state spaces.

2   In contrast to the Fordist-Keynesian project of alleviating spatial ine-qualities within the national territory through state action, the state spatial strategies associated with glocalizing state institutions have actively intensified intra-national socio-spatial polarization by promoting the reconcentration of economic assets, industrial capacities and infrastructural investments within the most powerful agglomerations. In this sense, glocalizing state spatial strategies are premised upon the assumption that intra-national uneven development may be continually instrumentalized as the *basis* for economic development rather than operating as a *barrier* to the latter.

The uneven development of regulation and the intensification of uneven development are thus important geographical-institutional dynamics within glocalizing states.

From contradictions to crisis-management

The new forms of uneven development unleashed through glocalization strategies are contradictory in the sense that they may hinder rather than support the processes of regulation and accumulation.

1   The increasing geographical differentiation of state regulatory activities may undermine the state's organizational coherence and functional unity, further exacerbating rather than resolving the crisis of spatial Keynesianism, leading in turn to serious governance failures and legitimation deficits (Painter and Goodwin, 1996).

2   The state's intensification of uneven spatial development within its own territory may seriously downgrade economic performance, as manifested in the overheating of the Southeast of England during the late 1980s (Peck and Tickell, 1995).

3   These dangers are enhanced still further by the zero-sum forms of interlocality competition that are promoted through glocalization strategies, which further destabilize an already uncertain economic environment at all spatial scales (Leitner and Sheppard, 1997).

In response to these dilemmas, a new politics of crisis-management appears to be emerging in which reformulated state projects and state strategies are being developed in order to address the regulatory deficits and structural contradictions associated with earlier modes of state spatial intervention. Particularly as of the late 1980s, when the contradictions of first-wave glocalization strategies became immediately apparent, this politics of crisis-management has arguably played an essential role in (re)moulding the institutional and geographical architectures of glocalizing states. As of this period, glocalization strategies began increasingly to encompass not only entrepreneurial approaches to urban development, but also a variety of flanking mechanisms intended to manage the tensions, conflicts and contradictions generated by earlier versions of such policies. Although these strategies of crisis-management have not prevented the aforementioned contradictions from being generated, they have generally entailed the establishment of various political-institutional mechanisms through which their most disruptive socio-economic consequences may be monitored, managed, and at least in principle, alleviated. This trend is exemplified in the recent reintroduction or rejuvenation of policies to address the problem of social exclusion in many European cities (Harloe, 2001).

The 'new regionalism' and the rescaling of glocalization strategies

The widespread proliferation of new regionally focused projects of state rescaling during the last decade may be understood in this context. The first wave of glocalization strategies focused predominantly upon the downscaling of formerly nationalized administrative capacities and accumulation strategies towards local tiers of state power. More recently, however, the regional or metropolitan scale has become a strategically important site for a major project to modify the geography of state regulatory activities throughout western Europe (Keating, 1997). From experiments in metropolitan governance and decentralized regional economic policy in Germany, Italy, France and the Netherlands to the Blairite project of establishing Regional Development Agencies in the UK, these developments have led many commentators to predict that a 'new regionalism' is superseding both the geographies of spatial Keynesianism *and* the forms of urban entrepreneurialism that emerged following the initial crisis of North Atlantic Fordism (for an overview, see MacLeod, 2000).

Against such arguments, the preceding discussion points towards a crisis-theoretical interpretation of these initiatives as an evolutionary modification of glocalizing state institutions in conjunction with their own immanent contradictions. Although the political-institutional content of contemporary regionalization strategies continues to be an object of intense contestation, they have been articulated thus far in two basic forms.

1   On the one hand, regionally focused strategies of state rescaling have frequently attempted to transpose entrepreneurial approaches to local econ-omic policy onto a regional scale, generally leading to a further intensification of uneven spatial development throughout each national territory. In this scenario, the contradictions of urban entrepreneurialism are to be resolved through the integration of local economies into larger, regionally configured territorial units, which are in turn to be promoted as integrated competitive locations for global and European capital investment. In this approach to regional state rescaling, the spatial selecti-vity of earlier glocalization strategies is modified in order to emphasize regions rather than localities; however, its basic politics of spatial reconcentration, zero-sum interterritorial competition and intensifying uneven development are maintained and unchecked.

2   On the other hand, many contemporary strategies of regionalization have attempted partially to countervail unfettered interlocality competition by promoting selected forms of spatial equalization *within* strategic regional institutional spaces. Although such initiatives generally do not significantly undermine uneven spatial development between regions, they can nonetheless be viewed as efforts to modify some of the disruptive aspects of first-wave glocalization strategies. Indeed, this aspect of regional state rescaling may be viewed as an attempt to reintroduce a downscaled form of spatial Keynesianism *within* the regulatory architecture of glocalizing states. The priority of promoting equalized, balanced growth is thus to be promoted at a regional scale within tightly delimited subnational zones, rather than throughout the entire national territory.

In short, both of the aforementioned, rescaled forms of crisis-management represent significant evolutionary modifications within glocalizing state apparatuses. While there is little evidence at the present time to suggest that either of these modified glocalization strategies will engender sustainable forms of economic regeneration in the medium-term, they are nonetheless likely to continue to intensify the geographical differentiation of state space and capital accumulation throughout western Europe.

CONCLUSION:

the new politics of uneven development

In this chapter, I have argued that a strategic-relational-spatial approach prov-ides a useful basis on which to explore the interplay between the rise of urban entrepreneurialism and processes of state spatial restructuring in western Europe. In this conceptualization, the rise of entrepreneurial approaches to urban governance has been intertwined with a broader redifferentiation and rescaling of national state spaces. Within the emergent, glocalized configuration of state spatiality, national governments have not simply transferred power downwards, but have attempted to institutionalize competitive relations between subnational administrative units as a means to position local and regional economies strategically within supranational circuits of capital. In this sense, even as the national scale of capital accumulation and state regulation has been decentred in recent decades, national states are attempting to retain control over major subnational spaces by integrating them within operationally rescaled, but still nationally coordinated, accumulation strategies. The concept of glocal-ization strategies is intended to provide a theoretical basis on which to grasp the increased strategic importance of urban and regional economic policies within this rescaled configuration of state spatiality.

As western European states seek to manage the tension between globalization and localization within their boundaries, the scalar organization of state space has become a direct object of socio-political contestation. The glocalization strategies analysed above represent a major expression of struggles to reorganize the geographies of state spatial regulation in strategic subnational spaces such as cities, city-regions and industrial districts. It appears unlikely, however, that these glocalization strategies will successfully establish a new structured coherence for sustainable capitalist growth. Instead, we appear to be witnessing processes of trial-and-error institutional restructuring, mediated primarily through *ad hoc* strategies of crisis-management and 'muddling-through'. In order to grasp such strategies, I have proposed a crisis-theoretical interpretation of recent regionally focused rescaling tendencies within glocalizing states. From this perspective, the contradictions unleashed through glocalization strategies are seen to provide an important impetus for their further evolution, in large part through the production of new scales of state spatial regulation. It is in the context of these emergent forms of crisis-management, I believe, that the much-discussed shift from a 'new localism' to a 'new region-alism' in many western European states must be understood. The evolutionary tendencies of rescaling within glocalizing state regimes therefore represent an important focal point for future research on entrepreneurial urban governance and state spatial restructuring in western Europe and beyond.

At the present time, the processes of globalization, European integration and EU-eastward enlargement have been dominated by neoliberal agendas that reinforce the entrepreneurial politics of inter-spatial competition described above. Meanwhile, the project of promoting territorial equalization within national or subnational political units is frequently dismissed as a luxury of a bygone 'golden age'. Yet, even as contemporary rescaling processes appear to close off some avenues of economic regulation, socio-spatial redistribution and democratic control, they may also establish new possibilities for the latter at other scales. For instance, the supranational institutional arenas associated with the EU may still provide a crucial mechanism through which progressive forces might mobilize political programmes designed to alleviate inequality, uneven development and unfettered market competition, this time at a still broader spatial scale than was thought possible during the era of high Fordism. It therefore remains to be seen whether contemporary dynamics of state rescaling will continue to be steered towards the perpetuation of neoliberal geographies of uneven development, or whether, perhaps through the very contradictions they unleash, they might be rechannelled to forge a negotiated political compromise at a European scale based upon substantive social and political priorities such as democracy, equality and diversity. Precisely because the institutional and scalar framework of European state space is in a period of profound flux, its future can be decided only through socio-political struggles, at a variety of scales, to rework the geographies of regulation and political mobilization. Under conditions such as these, the spatiality of state power has become the very object and stake of such struggles rather than a mere arena in which they unfold.

NOTES

<hr>

1   Jessop mentions a number of accumulation strategies, including Fordism, import-substitution and export promotion growth strategies in Latin America, the fascist notion of *Grossraumwirtschaft*, the West German *Modell Deutschland* and Thatcherism (Jessop, 1990: 201–2; Jessop et al., 1988).

2   This analysis is developed at greater length in Brenner (2003).

Chapter 13

GLOBAL PRODUCTION SYSTEMS AND EUROPEAN INTEGRATION:
DE-REGIONALIZING, RE-REGIONALIZING AND RE-SCALING PRODUCTION SYSTEMS IN EUROPE

Ray Hudson

INTRODUCTION:

globalization, international investment and global production strategies

As Peter Dicken (2000: 276, original italics) has recently put it, succinctly summarizing a distinctive characteristic of his sophisticated approach to understanding geographies of economies, *'places produce firms* while *firms produce places'*. In this chapter, I explore some of the ramifications of this perspective in terms of interrelationships between the globalization of production systems, the political-economic re-definition of Europe, processes of de-regionalization and re-regionalization of production linked to changes in corporate (dis)investment strategies and the changing character of places within Europe.

   Whilst the volume of international investment has increased within the latest phase of globalization – the latest and highest stage of capitalist development – it has displayed a distinctive macro-regional geography: the vast majority originates in north American, western Europe and Japan while the former two form the main destinations. There is abundant empirical evidence about flows of international investment into and out of Europe, primarily by major multinational and transnational corporations (MNCs and TNCs) but also by some small and medium-sized enterprises (SMEs). These flows form constitutive moments in creating global production systems and changing geographies of production within Europe. Whilst EU-based MNCs and TNCs seek investment opportunities in parts of Europe beyond their national territory and further afield outside Europe, major companies based in the USA and Japan in particular have invested within Europe. Whilst the volume of investment flows from Japan and other parts of southeast Asia rose sharply in the 1980s and 1990s (Dicken at al., 1997b), the stock of foreign direct investment (FDI) remains heavily dominated by US-based companies, which have a long history of investment in Europe (Dicken, 1998a). Moreover, Europe continues to be a key destination for FDI by US-based corporations (Deloitte Research, 2001). Furthermore, the balance between 'green field' investment in manufacturing and investment via mergers and acquisitions (M&A) – which now account for 75 percent of international investment (Nitzan, 2001) – has significantly shifted and is often linked to unprecedented growth in the number of strategic alliances and other forms of cooperative linkages between major firms.[1]

This complex pattern of flows, mediated through corporate organizational structures, raises important theoretical, political and practical questions, which may be summarized as follows:

1   How do the changing geographies of investment and production within Europe relate to processes of Europeanization and globalization of production systems and to processes of European political-economic integration and expansion, eastwards in particular?

2   How do MNC inward investment activities create new forms of cross-regional and cross-national integration within Europe and link parts of Europe with other parts of the world in global production systems that are both intra-corporate and inter-corporate?

3   What forces underlie the (non)coincidence of corporate and territorial development strategies within Europe, as articulated at EU, national and sub-national scales? How is this expressed in processes of re-regionalization and de-regionalization?

4   How can territorially defined political subjects (the EU, national states, regions or cities) in Europe 'capture', if only temporarily, particular activities and parts of corporate global value-creation chains?

5   Which places capture which activities? To what extent can public policies alter the characteristics of places to make them more attractive to particular sorts of inward investment?

RE-REGIONALIZING AND EMBEDDING PRODUCTION VIA NEW FORMS OF MULTINATIONAL AND TRANSNATIONAL INVESTMENT?

Growing globalization of the economy can be linked to the transition from MNCs to TNCs.[2] These changes in corporate form and strategy reflect an on-going search for more effective ways of creating new forms of uneven development and exploiting existing ones in pursuit of competitive advantage and profit.

As Dicken (1998a: 201–41) has emphasized, TNCs and MNCs invest for diverse reasons in varied locations, thereby linking them in often complex ways within their intra-corporate but nonetheless global systems of production. These can be summarized under three headings related to variations in corporate competitive strategy:

1   accessing markets, often protected via tariff barriers, and taking advantage of economies of scale and differentiated consumer markets;

2   lowering production costs via finding low cost locations and/or achieving economies of scale and scope; and

3   acquiring knowledge and other assets that allow radical change in techno-organizational production paradigms and in forms of competition, temporarily securing competitive advantages that translate into increasing market share for existing products or a monopoly position in markets for new products. This radically re-defines the forms, dimensions and scope of competition.

Reflecting these three qualitatively different forms of corporate strategy, TNCs create linkages and produce economic integration across national and supranational boundaries in three analytically distinct ways (Dunning, 2000). The first two are asset exploiting:[3] firstly, horizontal integration with the same product (or locally customized versions thereof) supplied by the same firm in different countries; and secondly, vertical integration with different stages of the production process and value creation chain for a given commodity undertaken in different locations within an evolving international division of labour. Much inward investment in manufacturing in Europe involves these forms of integration driven by asset-exploiting corporate competitive strategies.

While there has been some inward investment by manufacturing firms in activities such as European headquarter offices and R&D in core regions, to date, it has generally been limited in extent. Such activities typically remain deeply embedded in home countries, with this spatial concentration of R&D in the national base of the innovating firm memorably characterized as 'an important case of non-globalization' (Pavitt and Patel, 1991; see also Gertler, in this volume). For example, there is a significant difference between the level of R&D investment by indigenous and foreign-owned companies (on average 5–7 per cent and 2–2.5 per cent, respectively). While there are variations within both groups, with several overseas companies investing heavily in R&D, 'these are the exceptions rather than the rule' (John Dodd, the UK Federation of Electronics Industries, quoted in Cane and Nicholson, 2001).

While these exceptions are important in relation to claims about new forms of more 'embedded' factories, most FDI in manufacturing, both in 'green field' factories and via acquisition of existing plants and companies, has been in routine production activities either to guarantee access to European markets for key consumer goods (automobiles, 'white goods') and/or to find lower cost production locations, often exporting the resultant output beyond the EU's boundaries. New green field factories have typically been established in peripheral locations characterized by high unemployment, abundant cheap and often highly skilled manual labour and generous state financial subsidies. In electronics in the UK, for example, 'most foreign-owned companies are essenti-ally offshore manufacturing sites' (Dodd, quoted in Cane and Nicholson, 2001). In many ways, such plants define the archetypical 'branch plant' economy with low levels of local sourcing and an absence of advanced R&D. Such branch plants are located both in the peripheries of Europe and at the extremities of global chains of corporate command and control, vulnerable to closure because of decisions taken in distant locations. Moreover, continuing branch plant investment in Europe's peripheries bears a close resemblance to the specifica-tion of a 'global outpost', especially in regions within or just beyond but soon to be within the eastern boundaries of the EU (Hudson, 2002).

However, there are claims that radically new forms of branch plant investment began to be made in Europe from the 1980s in response to the crisis of Taylorist mass production and corporate search for new models of more flexible high volume production (FHVP), combining the benefits of economies of scale and scope. As a consequence of the adoption of 'just-in-time' principles of production, this has often been associated with regional re-concentration of production in particular industries (archetypically automobiles) and 'embed-ding' in particular places. This is seen as denoting a greater permanence of investment and commitment to place – for example, via investment in R&D – than in the archetypical 'global outpost'. These processes of re-regionalization and embedding come about because lead manufacturers develop component supply chains in the surrounding region and secure provision of required 'hard' and 'soft' infrastructures – for example, via state investments in transport, telecommunications and IT networks and developing capacity for training required labour, often in collaboration with state organizations. However, 'just-in-time' is not necessarily equivalent to 'in-one-place'. As a result, such new 'embedded' production complexes in Europe are also linked, often over great distances, with other factories both as sources of component inputs and markets for outputs within complex and distanciated globalized production systems.

Rather than regarding 'global outposts' and ' embedded branch plants' as dichotomous alternatives, however, it is useful to conceptualize them as opposites ends of a continuum, with particular plants tending more towards one pole or the other in a variety of hybrid forms. These conjoin company and territory in a variety of ways. The specific form depends upon the interplay of corporate strategy and the specific reasons for establishing (or acquiring) a particular plant in a particular time/space and the strategies, power and competencies of relevant economic development organizations. Drawing on Kindleberger (1969) and his concept of the 'enclave economy', Phelps and MacKinnon (2000) conceptualize such production systems as industrial enclaves, exhibiting more connection to the surrounding region than 'global outposts' but less than the deep and far-reaching intra-regional connections, commitments and integration of 'embedded branch plants'. There is certainly evidence of continuing investment in 'global outposts'; there is also evidence of branch plant investment, which approximates more closely to the specifications of 'embeddedness', in factories mandated to perform a wider range of functions that those 'traditionally' associated with 'branch plant' investment. For example Fiat's Melfi plant and the Nissan plant at Sunderland both have several of the attributes associated with 'embedded investment' – for example, intra-regional supply chains and close links with local economic development and training institutions.[4] Many MNC plants in the IT sector in Ireland began as simple assembly operations, but subsequently added R&D and product development functions (Brown, 2001). Furthermore, many foreign manufacturing plants incorporate both upstream (R&D) and downstream (sales and marketing) functions, as well as procuring inputs via intra-regional supply chains and linking with local institutions within the state and civil society.

In such instances, therefore, there *may* be evidence of the emergence of regional production complexes with functional integration between co-located activities within and beyond the boundaries of the firm. Moreover, the growing recognition of the importance of incremental innovation and learning by doing within communities of practice suggests that branch plants within hybrid forms of 'enclave' may become repositories of practical know-how that provide localized competitive advantage (Morgan, 1997) and also a cognitive resource that can be deployed more widely within the firm. However, it is necessary to distinguish carefully between the presence of particular functions in a plant and the ways in which these integrate it into the surrounding regional economy and the (global) corporate production system. Co-location of functions does not *necessarily* equate to the presence of a regionally integrated production system. For example, there may be R&D or marketing activities, but these activities may be linked more closely to manufacturing activities in *other* locations. Such branch plant investment approximates more closely to the specifications of hybrid forms of 'enclave' rather of deep 'embeddedness'. These contested claims about 'embedded' production complexes and the variety of forms of 'enclave' branch plant investment that have emerged raise questions about the ways in which global production systems are becoming 'fixed' in Europe, as major corporations both use existing spatial differences and create new forms of spatial differentiation and link locations in complex ways as constitutive and integral elements within their global production strategies.

The first two forms of cross-border integration discussed above result from corporate strategies of asset-exploiting weak competition within a given techno-organizational paradigm. The third type of integration results from corporate strategies of asset-augmenting Schumpeterian strong competition that seek to redefine production paradigms and possibilities for creating and appropriating surplus value. It involves seeking to capture 'local' knowledges and enhance R&D capabilities, often via acquiring or accessing assets created abroad to protect or enhance competitive advantage and core competencies. As such, growing emphasis upon asset-augmenting investment is linked to important changes in the form and 'territoriality' of transnational investment. In particular, it is linked in various ways with a tendency towards 're-regionalization' of production in 'sticky places' (Markusen, 1996) and with corporate strategies that use M&A to find the locally constituted intellectual assets that such places create and exploit them within globalized production systems. These assets include 'local knowledge', tacit knowledge (firm-specific and place-specific) about production and new 'scientific' knowledge produced within the R&D laboratories of other companies and organizations about new, potentially revolu-tionary 'market disturbing' products and processes.

While acquisition is often a very cost-effective way of obtaining knowl-edge about products and processes, companies acquiring others in search of knowledge also acquire material and other assets. They must decide which of the activities of the newly acquired company are relevant to their own core competencies and, conversely, which activities are tangential and which product lines have surplus capacity and either close them down or divest them to others. Divestment, capacity reduction and 'exit' from 'non-core' activities are often key elements in post-acquisition rationalizations and have had marked effects on geographies of production in Europe.

There has also been a growing tendency to enter strategic alliances – on a multiple basis, with the same firm typically involved in several – amongst the major corporations, as they use these to acquire new knowledge, to learn new ways of producing via long-term strategic links with other major companies as well as with smaller and structurally less powerful firms. Such links are said to depend upon trust and 'trust' can undoubtedly offer considerable advantages in reducing time and effort, risk and uncertainty and facilitating learning via enhancing the quality and quantity of information flows (Morgan, 2001). However, trust also carries costs – for example there are dangers of 'lock-in'. There are definite limits to cooperative behaviour and trust because of the structural constraints that arise from the inherently antagonistic and competitive character of capitalist economic relations (Hudson, 2001a). Consequently, acqu-iring knowledge and learning in this way poses severe challenges for companies (see Gertler and Amin, in this volume).

Globalized production systems are thus constituted via a complex dialectic of competition and cooperation, distanciation and localization/regionalization. As Dicken et al. (1994: 30) rightly emphasize, a key diagnostic feature of the 'newly emerging organizational form' of the 'complex global firm' is an 'integrated network configuration and... capacity to develop flexible co-ordinating processes', both inside and outside the firm. Understanding the ways in which places within Europe fit into the changing geographies of globalizing production systems requires such an analytic point of departure. Equally, the distinctions between asset-exploiting and asset-augmenting motivations for FDI are analytic. Actual firms operating in real historical time/space typically deploy them simultaneously and in varying combinations. Understanding changing geographies of economies within Europe requires recognition of this.

HOLLOWING OUT INDUSTRIAL DISTRICTS AND RELATED FORMS OF REGIONALIZED PRODUCTION SYSTEMS

The integration of Europe into global production systems is not, however, simply a consequence of the activities of big companies. Indeed, in other ways it is intimately linked to a transition in what were previously seen as very 'sticky' places in Europe, notably industrial districts such as those of the Third Italy. These have become less cohesive as the socio-economic glue of beliefs and material practices (rather than simply something 'in the air') that previously held them together strongly in place has, to varying degrees, dissolved.

From the early 1980s, the larger or leading firms in quintessential Italian clothing industrial districts such as Carpi and Prato initiated a far-reaching process of selective de-localization of labour-intensive and unskilled stages of production. Conversely, they increasingly concentrated upon high quality products and those stages of production requiring skilled labour. More import-antly, they focused upon design, marketing and brand development, as well as key HQ strategic functions, activities less sensitive to labour costs. Such de-localization tendencies also reflected growing resistance by women, children and marginalized workers to 'super-exploitation', in strong contrast to the dominant image of these districts as characterized by egalitarian, progressive industrial relations (Hadjimichalis and Papamichos, 1990). Similar processes of 'hollowing out' production to surrounding localities with abundant cheap labour occurred elsewhere in southern Europe.[5] For example, in the 1980s there was a relocation of production to areas around the town of Kastoria in northern Greece, the one authentic industrial district in Greece, producing expensive clothing from imported fur (Hadjimichalis, 1998).

The 'hollowing out' of industrial districts can be a complex process, involving re-locating production beyond the national territory, as the example of Benetton illustrates. Benetton emerged as a major clothing company because of a complex combination of marketing, organizational and process innovations. This encompassed creating a new global product image, a refined just-in-time production system incorporating both out-sourcing and process innovation,[6] and a risk-minimizing strategy of franchised outlets in over 100 countries, while retaining key control, design and marketing functions in Treviso, northern Italy (Crewe and Lowe, 1996). Benetton was initially seen as an anomaly, a rare exception. But increasingly other districts unravelled in similar ways. The boundaries of clothing industrial districts became more permeable because of the emergence of powerful 'lead' firms or *gruppi*, a result of organic growth or, more often, of M&A activity among local firms and the entry of externally-owned larger firms that came to play dominant roles and shape local growth and development (Whitford, 2001). This created more complex structures of owner-ship and changed relationships between firms. Previously egalitarian horizontal relations have become much more hierarchical and asymmetric, radically resha-ping the anatomy of power relations between firms sharing the same location. In addition, lead firms establish relationships with suppliers and subcontractors beyond the boundaries of the district and fracture the former territorially bounded coherence and integrity of the production system within the district. Consequently, there has been a growing tendency to recast these places as nodes in wider European or global corporate production systems.

In addition, however, such changes were also a consequence of develop-ments beyond the boundaries of the districts. These changes enabled shifts in corporate behaviour and indeed were necessary conditions for them. In particu-lar, they were a result of the removal of previous constraints because of broader geopolitical changes in Europe. The opening up of the formerly forbidden territories of central and eastern Europe (CEE) to capitalist investment has been especially important in this regard. Two examples illustrate the point. Herning-Ikast in Jutland developed as a sophisticated industrial district, based on the production of high-value added woollen clothing (Maskell et al., 1998). During the 1990s, however, production tasks, especially the most labour intensive ones, were subcontracted to firms in Poland, while the key decision-making, marketing and design functions remained in Herning-Ikast. Around the same time, in the metal working industrial district around Brescia in northern Italy, basic steel making activities were increasingly subcontracted to firms in Bulgaria and Rumania, while keeping key design, marketing and decision-making functions in and around Brescia. In both cases, there was a 'hollowing out' of the production structure of the industrial district in response to newly available locational opportunities driven by increasingly fierce pressure on global product markets. In this way, indigenous European SMEs became enmeshed in evolving global production systems. In addition, other small firms in many parts of Europe have become tied into global systems via the growth of subcontracting and new forms of collaborative inter-firm relations. It is, how-ever, important to emphasize that collaboration in this context does not mean equality of power between partners in such agreements. As a result, many small firms in Europe came to occupy vulnerable and precarious market niches within global production structures rather than within local or national markets, often subject to fierce price competition from locations outside Europe.[7]

However, there are also counter-tendencies that involve re-concentration of production in these same industries in other parts of Europe. While long established industrial districts were being hollowed out and reorganized, new clothing clusters were emerging elsewhere. In CEE, political change provided opportunities for a degree of re-regionalization of clothing production as new clusters evolved, incorporating innovative forms of inter-firm relations, linked into local 'lead' production firms and in turn into export markets in western Europe.

Major European retailers have sought to supply from within Europe rather than lower-cost non-European locations, linked to moves to smaller batch production and the need for rapid response to market changes (Crewe and Lowe, 1996). They are reorganizing supply chains, often involving sourcing from locations in CEE in which labour is more expensive than in – say – India – but much less expensive than in western Europe, while transport times and communication problems are much less between eastern and western Europe than between western Europe and distant parts of Asia (Dunford et al., 2001). Similar processes were evident in parts of southern Europe, but focusing more upon specialized niche production. For example, in the rural areas of the Ave Valley in northern Portugal, clusters of clothing producers increasingly focused upon small batch production, manufacturing products for which the main modality of competition is quality rather than price (Thiel et al., 2000).

Companies located in other forms of regionalized production systems within Europe are also increasingly seeking to 'hollow out' routine production activities in response to growing competitive pressures. One such region is Baden-Württemberg in Germany. Although often classified as an industrial district (for example, see Scott, 1988a), there are significant differences in industrial culture and structure that differentiate it sharply from the canonical industrial districts discussed above. In particular, in Baden-Württemberg, an hierarchically organized production system evolved around large 'lead' firms (some indigenous German, others inward investing TNCs) and very large factories (for example, the Singelfingen automobile factory of Daimler Benz employed over 40,000 people as recently as the early 1990s) at the centre of regionalized supply structures involving multiple sourcing within the region. These component supply companies were themselves often substantial. Some, such as Robert Bosch, are major TNCs. Even medium-sized component suppliers would typically have anything up to 2,000 employees in their factories.

Increasingly, however, this regionalized production structure has become uncompetitive, rigid and inflexible. As a result, Daimler-Benz, Audi, Robert Bosch, IBM, Hewlett Packard, SONY – all large TNCs based in Baden-Württemberg – 'have in fact begun engaging in [the] process of killing off the old regional division of labour that they were embedded in. This has involved a sharp winnowing of the number of suppliers that firms engage with not simply in general but also within the region itself' (Herrigel, 2000: 296). Collaborative production in autos or electronics blurs the boundaries between firms, but it also involves significantly fewer firm boundaries: 'multiple sourcing has been replaced with long-term contracting with smaller numbers of intimate firms. These firms can be local firms, *but they need not be*' (Herrigel, 2000: 296; original italics). Consistent with this process of hollowing out and de-region-alization, the large firms have shown a willingness to invite in foreign expert collaborative suppliers, such as the Canadian firm Magna, to establish green field operations in the region. Furthermore, expert collaborator firms can be accessed from their locations elsewhere in Germany and in Europe (including CEE), without requiring relocation into Baden-Württemberg. Thus, 're-regionalisation occurs but it does not have to occur within the old geographic boundaries of the traditional district' (Herrigel, 2000: 297).

In summary, there are increasing tendencies to 'hollow out' the production structures of regions that until recently were regarded as archetypical examples of regionalized production systems within Europe. Responding to growing competitive pressures and taking advantage of opportunities offered by technol-ogical and geopolitical changes, established or emergent 'lead' firms in these regions are reorganizing the socio-spatial structures of their production systems, decentralizing much of the materials transformation stages of production to other locations. As a result, the home regions of these lead firms are increas-ingly becoming systems integrators, sites of key decision-making, design and marketing within wider, spatially dispersed production systems that link a variety of other firms and places. At the same time, the reorganization of produc-tion may lead to a rescaling of production, with the emergence of new clusters of firms in other locations, a process of re-regionalization and possibly rescaling of production as changes in corporate strategy create new scales of spatial organization.

POLITICAL-ECONOMIC INTEGRATION IN EUROPE AND THE WIDENING AND DEEPENING OF THE EU

The creation of a common economic space over Europe forms one moment in a more general process of global macro-regionalization, with the creation of free-trade areas in other parts of the world (such as NAFTA and MERCOSUR). Processes of political integration have produced a still evolving multi-scalar system of governance and regulation in Europe, especially the EU. The 'reorganization' of the state involves a triple process of de-nationalization (hollowing out), de-statization of the political system and the international-ization of policy regimes (Jessop, 1997; Brenner, in this volume). De-statization is important in changing the balance between the institutions of civil society and the state in governance, but is not as relevant here as de-nationalization and internationalization of policy regimes. Furthermore, there is growing complexity and, in some respects, new forms of uneven development in EU-level policy regimes. The deepening of the EU, notably via the Single European Market programme and the creation of the euro single currency, has created important divisions between those national states that are members of Euro-land and those that are not, both within and outside the boundaries of the EU. These processes of reorganization are changing the relations and balance of power between different levels of the state within the EU. At the same time, especially following the collapse of state socialism in CEE and the subsequent initiation of processes of political and economic transformation there, the ongoing process of EU enlargement is adding to this complexity.[8]

While power, authority and competencies have to a degree been shifted 'up' from national to EU level; there have also been pressures to move decision-making power and resources to subnational levels of cities and regions, increasing the role of specifically regional and urban institutions in governance and regulation. However, there are also counter-tendencies that re-emphasize the significance of the national. Consequently, although the process of unbundling territoriality has gone further in the EU than elsewhere (Ruggie, 1993), national states in Europe are neither dying nor retiring, but they have merely shifted functions (Mann, 1993). The claim of 'neo-medievalists' that the national state is largely rendered redundant in Europe is unfounded and reports of 'the exaggerated death of the nation-state' are premature (Anderson, 1995). Instead, national states are taking on different roles within complex multi-scalar structures of regulation and governance. There are complex links between this emerging multi-scalar political system and processes of economic internationalization, both Europeanization and globalization. The latter are reshaping the economic structure and geography of Europe, while political subjects are seeking to influence international investment flows, both inward investment to and outward investment from the EU and its constituent national territories.

As an embryonic emergent 'super-state', the EU both encourages and seeks to resist processes of globalization. The EU has long sought to encourage the formation of 'Euro-champions', European TNCs that can compete in global markets with those of the US and Japan in particular. In part, it has done so via pursuing strategic industrial policies that seek to shape market conditions confronting all firms in a sector, promoting inter-firm cooperation in order to ensure competitive success. For example, RTD initiatives such as ESPRIT aimed to encourage inter-firm collaboration in high-tech industries over pre-competitive technological innovation. The EU has also encouraged the emergence of Euro-champions via its permissive and non-interventionist attitude towards acquisitions of EU companies by other EU companies and to intra-EU mergers.[9] Even so, it has had limited success in encouraging and facil-itating successful cross-border M&A within Europe. In part, this is precisely because of the enduring significance of national differences in industrial cultures and attitudes towards activity. Some cultures of capitalism encourage M&A (notably the Anglo-American model), while others are, or until recently have been, less receptive to such corporate behaviour, especially in cases in which this involves unwelcome 'hostile' acquisitions (the German and Swedish models, for example).

Moreover, while the EU Commission encouraged the emergence of Euro-champions, many national governments have aimed to create 'national champions' to protect national interests in strategically significant industries, such as those in areas of emergent new technologies and defence (for example, the French and Italian governments).[10] However, the EU Commission has actively sought to discourage or prevent M&A intended to create national champions (Guerrera, 2001).[11] This (as with the inability to agree regulations for cross-border M&A) is indicative of a continuing struggle between national states and the Commission over the architecture of multi-level governance within the EU territory and as to where regulatory authority and competence should reside in Europe. It is also a struggle over the preferred corporate anatomy of the EU. One, perhaps unintended, consequence of these varying political priorities and strategies towards M&A has been to enhance the role of inward investment from outside the EU in such activity. As a result, it has enhanced the significance of TNCs based outside the EU in shaping economic development trajectories within Europe. Consequently, while the Single Europ-ean Market led to a surge in M&A activity within the EU in the 1980s and early 1990s, this mainly involved inward investment by TNCs based outside the EU. As a result, the EU has sought to influence the intra-EU geography of international inward investment via both its own industrial and regional policies and regulation of national industry and regional aids, including setting limits to national aids and limiting the extent of competitive bidding between countries and regions for major projects.

As noted above, national governments remain important actors in processes of economic governance and regulation, including the attraction of FDI from abroad. Virtually every European country has considered FDI to be sufficiently significant to warrant a set of specific polices and organizations focused on its attraction. Furthermore, there has been a growing sophistication in policy design and delivery at the national level. However, national states in Europe vary in the extent to which they wish to embrace processes of international competition and global markets as economic steering mechanisms and seek to attract FDI. This can be related to the well-known persistence of national distinctiveness in forms of capitalism within Europe (Hudson and Williams, 1999). Broadly speaking, national states can be divided into three groups in terms of their attraction of FDI (Brown and Raines, 2000: 436–40). First, those with long-standing policies to maximize the attraction of FDI (e.g. the UK and Ireland). Second, those that have a continuing mistrust of inward investment because of fears of losing national control of key economic sectors or of indigenous firms losing out in competition with inward investors (e.g. Germany and Italy). Third, countries that were formerly hostile to FDI, but are becoming, or have become, much more favourably disposed towards it (e.g. France and the countries of the Mediterranean and Scandinavia and, above all, after 1989, those of CEE). However, while national variation and specificity remain, there is increasing convergence on a broadly neoliberal UK/USA model that encourages privatization of sectors of the economy previously reserved for the state and public sector and creates space for international inward investment in this and other ways (see Tickell and Peck, in this volume). This convergence is a conseq-uence of the process of deepening economic integration in the EU and one that is increasingly driven by concerns with shareholder value rather than equity between the various social partners and stakeholders as a consequence of the effects of globalization of financial markets.

At subnational levels, cities and regions in Europe have engaged in an increasingly generalized place-marketing competition for FDI from other countries within and beyond the EU, especially the US and Japan. Indeed, the attraction of FDI has increasingly been devolved to local and regional levels as an integral part of the process of developing more sophisticated and sensitive national policies. Local and regional authorities within Europe have sought to construct economic development strategies around varying types of FDI, depending upon their location and the characteristics of their areas. Choice of strategy often reflects the character and location of the place, linked to an evolving spatial division of labour within Europe that is increasingly differenti-ating qualitatively between places in terms of their location in circuits of value creation, appropriation and realization (Hudson,

2001b). The key issue is what sorts of activities can be attracted to and – for a time – held down in what sorts of places? For example, within the UK, it was feasible to seek to redevelop the London Docklands around international investment in high-order financial and business services. In contrast, in peripheral places within Europe, the focus of regeneration policies has been manufacturing branch plants (preferably embed-ded performance plants at the heart of new clusters rather than Taylorized global outposts, but if the latter are all that are on offer, welcoming them nonetheless), or the Taylorized 'back office' service activities of financial and business services. However, both the quantitative and qualitative employment creation effects of such investments in peripheral regions are very limited (Hudson, 1999; 2002), not least because the prevailing high levels of unemployment in these regions enable companies to be extremely selective in their recruitment policies.

The institutional character and performance of places, their 'soft' as well as their 'hard' infrastructures have been increasing emphasized in both the acade-mic and policy literatures about relations between FDI and regional economic performance. This has been linked with growing concerns to use inward investment to stimulate new 'clusters' and in this way re-regionalize production and revivify regional economies. Consequently, the form and content of local and regional development policies appropriate to first attract and capture, then – as the flows of FDI into the EU and applicant states in CEE have declined – retain and nurture inward investment projects via 'after-care' policies has increasingly come under scrutiny. Indeed, the emphasis upon 'after-care' may lead to local and regional development organizations being 'captured' by inward investing TNCs, which can exert considerable influence over their actions and policies (Phelps et al., 1998). In such cases, the balance of corporate and territorial developmental interests tilts sharply in the direction of the former to the detriment of the latter.

CONCLUSION

Geographies of economies within Europe have undoubtedly changed – and continue to change – in important ways, with simultaneous tendencies towards de-regionalization and re-regionalization of production often at new spatial scales. These changes reflect the ways in which political actors seek to capture corporate investments and companies seek to use locations within Europe as part of their evolving strategies to develop more globalized production systems. This involves inward investment to and outward investment from Europe, as well as the creation of a variety of inter-firm linkages ranging from out-sourcing strategies to the creation of long-term strategic alliances. Increasingly these wider linkages are driven by the need to acquire knowledge and a variety of intangible assets and create complex ecologies of learning to enable companies to do so rather than merely access to lower cost production sites or to major markets. Consequently, there has been a switch in emphasis to asset-augmentation in corporate strategies, although asset-exploitation remains a major influence shaping many corporate geographies of production.

These changes in forms of material and knowledge production and the resultant re-definition of intra-European geographies of production are also connected in significant ways to the evolution of the EU. This is the case both in terms of its expansion to incorporate further parts of Europe (and arguably areas beyond Europe as conventionally understood) and in deepening processes of political-economic integration within this expanded space. As yet, deepening integration has not led to convergence in national economic structures or to increasing regional specialization (Geroski and Gugler, 2001). However, these processes are still in their early stages and a corollary of their maturation could well be the creation of new forms of uneven development as places are increasingly and qualitatively differentiated in terms of their position in wider European and global production systems. As a result, in some places there would be growing convergence between corporate and territorial development interests, but in many others there would be growing divergence between these two sets of interests. Consequently, there may well be the simultaneous creation of virtuous circles of development for some places but vicious cycles of decline for others, as the contours of the map of economic well-being in Europe become increasingly sharply delineated. The resultant widening of the map of inequality could have potential explosive implications for social cohesion within Europe. This in turn may create major political uncertainties as to the place that many parts of Europe can occupy within globalizing production systems as the cycle of uneven development is reproduced in increasingly sharp form. Understanding these processes of combined and uneven corporate and territorial development, as the emergence of global production systems runs up against the necessary territoriality of production, poses significant theoretical challenges for economic geographers.

ACKNOWLEDGEMENT

This chapter draws in part on two recently completed ESRC-funded research projects: 'Social Exclusion or Flexible Adaptation? Coalfields in a Period of Economic Transformation' was carried out with Professor Huw Beynon and examined inward investment as part of coalfield regeneration strategies in the UK. 'Regional Economic Performance, Governance and Cohesion in an Enlarged Europe' is part of ESRC's 'One Europe or Several?' programme and is focused upon five sectors (automobiles, clothing, chemicals, food and steel) in eight regions of the EU and current EU applicant states (two each in Italy, Poland, Slovakia and the UK). I would like to acknowledge the contributions of other colleagues working on this project – David Dornisch, Michael Dunford, Jane Hardy, Bill Heywood, Al Rainnie, David Sadler and Adrian Smith. It also draws on research carried out in Denmark, Germany and Greece during the 1980s and 1990s.

NOTES

1   A considerable amount of inward FDI into the EU has been in service sector activities such as banks and financial services that have distinctive intra-EU geographies and are different to those of investment in manufacturing (Dicken et al., 1997a; Hudson, 1999). However, because of constraints of space in this chapter, I focus upon the globalization of material production and manufacturing while recognizing that the increasingly complex social division of labour and the growth of networks and of out-sourcing is rendering redundant the conventional statistical distinction between 'manufacturing' and 'services' (Hudson, 2001a). Indeed, much FDI 'service' investment within the EU has been by 'manufacturing' companies – for example in distribution functions.

2   Although there are very few genuinely TNCs, many exhibit some tendencies towards transnationalization (Allen and Thompson, 1997; see also Dicken, in this volume). It may be more accurate to refer to a transition from multinational to transnational strategies within corporations of varying sizes.

3   The third type of strategy, asset-augmenting, is discussed below.

4   However, following the strategic alliance between Nissan and Renault in 1999, there have been important changes that have loosened the degree of regional embedding of the Nissan plant – for example, purchasing decisions have been removed from plant level and centralized in Paris (Hudson, 2002).

5   While there are many concentrations – or clusters – of small firms specializing in specific products in southern Europe (Garofoli, 2002), most of these are not organized around the social relations of production that characterize industrial districts (see Malmberg, in this volume).

6   In 1972, Benetton introduced in-house dyeing at the final stage of production, crucially allowing piece (rather than batch) dyeing and so the dyeing of individual items to order.

7   Often the location of these firms in wider production structures has been decisively shaped by trade policies. For example, the outward processing trade (OPT) regime established as part of the Multi-Fibre Arrangement (introduced in 1974 under the aegis of GATT) was critical in shaping subcontracting relationships between small clothing producers in southern, central and eastern Europe and major clothing retailers in western Europe.

8   A corollary of this is an active debate as to where the boundaries of Europe are to be drawn, who is to be permitted to move across these boundaries and who is to be defined as 'European' (Hudson, 2000).

9   Since EU regulators began to scrutinize merger and acquisition proposals in 1990, only 16 have been blocked (Guerrera and Mallet, 2001).

10   One consequence of this is that the EU has been unable to establish agreed rules for cross-border mergers and acquisitions. Opposition of some national governments, especially Germany, has impeded moves towards the creation of European champions (Mann, 2001).

11   The Commission effectively seeks to force companies in national markets with highly concentrated sectors to acquire or merge elsewhere in Europe and create Euro-champions that conform to its normative vision of the corporate anatomy of the EU.

REFERENCES

Abo, T. (ed.) 1994. Hybrid Factory. New York: Oxford University Press.

Abo, T. 1996. The Japanese production system. In R. Boyer and D. Drache (eds), States Against Markets. London: Routledge, 136–54.

Agamben, G. 1998. Means Without End. Minneapolis, MN: University of Minnesota Press.

Agnew, J. 1994. 'The territorial trap'. Review of International Political Economy, 1: 53–80.

Albert, M. 1993. Capitalism Against Capitalism. London: Whurr.

Albrow, M. 1996. The Global Age. Cambridge: Polity Press.

Allen, J. and Thompson, G. 1997. Think global, then think again – economic globaliz-ation in context. Area, 29: 213–27.

Almeida, P. and Kogut, B. 1999. 'Localization of knowledge and the mobility of engineers in regional networks'. Management Science, 45: 905–17.

Amato, J.A. 2000. Dust. Berkeley, CA: University of California Press.

Amin, A. 1997. 'Placing globalization'. Theory Culture and Society, 14: 123–37.

Amin, A. 2000. 'Organisational Learning Through Communities of Practice'. Paper presented at the Workshop on The Firm in Economic Geography, University of Portsmouth, UK, 9–11 March.

Amin, A. 2001. 'Globalization: geographical aspects'. In N. Smelser and P.B. Baltes (eds), International Encyclopedia of the Social and Behavioural Sciences. Amsterdam: Elsevier Science, 6271–7.

Amin, A. and Cohendet, P. 1999. 'Learning and adaptation in decentralised business net-works'. Environment and Planning D, 17: 87–104.

Amin, A. and Cohendet, P. 2000. 'Organisational learning and governance through embedded practices'. Journal of Management and Governance, 4: 93–116.

Amin, A. and Cohendet, P. (forthcoming) Organizing Knowledge. Oxford: Oxford University Press.

Amin, A. and Thrift, N.J. 1992. 'Neo-marshallian nodes in global networks'. International Journal of Urban and Regional Research, 16: 571–87.

Amin, A. and Thrift, N.J. 2002. Cities. Cambridge: Polity Press.

Amsden, A. 2001. The Rise of the Rest. Oxford: Oxford University Press.

Anderson, J. 1995. The exaggerated death of the nation state. In J. Anderson, C. Brook and A. Cochrane (eds), A Global World? Oxford: Oxford University Press, 65–112.

Anderson, S. and Cavanagh, J. 2000. Top 200: The Rise of Corporate Global Power. Washington, DC: Institute for Policy Studies.

Andersson, U. and Forsgren, M. 2000. 'In search of excellence: network embedddness and subsidiary roles in multinational corporations'. Management International Review, 40: 329–50.

Andersson, U., Forsgren, M. and Holm, U. 2001. 'Subsidiary embeddedness and competence development in MNCs'. Organization Studies, 22: 1013–34.

Angel, D.P. and Engstrom, J. 1995. 'Manufacturing systems and technological change'. Economic Geography, 71: 79–102.

Annan, K. 2001. Laying the foundation for a fair and free world trade system. In G. Sampson (ed.), The Role of the World Trade Organization in Global Governance. Tokyo: United Nations University, 19–28.

Appadurai, A. 1986. Introduction: commodities and the politics of value. In A. Appadurai (ed.), The Social Life of Things. Cambridge: Cambridge University Press, 3–63.

Appadurai, A. 1996. Modernity at Large. London: University of Minnesota Press.

Argyris, C. and Schon, D. 1978. Organizational Learning. London: Addison Wesley.

Arrighetti, A., Bachmann, R. and Deakin, S. 1997. 'Contract law, social norms and inter-firm cooperation'. Cambridge Journal of Economics, 21: 171–95.

Arronson, S.A. 2001. Taking Trade to the Streets. Ann Arbor: University of Michigan Press.

Arthurs, H.W. 2000. The hollowing out of corporate Canada? In J. Jenson and B. Santos (eds), Globalizing Institutions. Aldershot: Ashgate, 29–51.

Asheim, B.T. 1996. 'Industrial districts as "learning regions"'. European Planning Studies, 4: 379–400.

Audretsch, D.B. and Feldman, M.P. 1996. 'Knowledge spillovers and the geography of innovation and production'. American Economic Review, 86: 630–40.

Baldwin, R. 1970. Non Tariff Distortions of International Trade. Washington, DC: Brookings Institutions.

Baldwin, R. 2001. Regulatory protectionism, developing nations, and a two-tier world trade system. In S.M. Collins and D. Rodrik (eds), Brookings Trade Forum 2000. Washington, DC: Brookings Institutions.

Ball, G. 1967. 'Cosmocracy'. Columbia Journal of World Business, 2/6: 25–30.

Baptista, R. and Swann, P. 1996. The Dynamics of Industrial Clusters. Centre for Business Strategy Working Paper 165, London Business School.

Baptista, R. and Swann, P. 1998. 'Do firms in clusters innovate more?' Research Policy, 27: 527–42.

Barkema, H.G. and Vermeulen, F. 1998. 'International expansion through start-up or acquisition'. Academy of Management Journal, 41: 7–26.

Barnes, T.J. 2001. 'Retheorizing economic geography'. Annals of the Association of American Geographers, 91: 546–65.

Barnet, R.J. and Muller, R.E. 1974. Global Reach. London: Jonathan Cape.

Bartlett, C.A. and Ghoshal, S. 1989. Managing Across Borders. London: Century Business.

Bartlett, C.A. and Ghoshal, S. 1995. Transnational Management. Second Edition, Chicago: Irwin.

Basch, L., Glick Schiller, N. and Blanc-Szanton, C. 1994. Nations Unbound. Langhorne, PA: Gordon and Breach.

Beaverstock, J. 1996. 'Migration, knowledge, and social interaction: expatriate labour within investment banks'. Area, 28: 459–70.

Bennell, P. 1997. Privatisation in sub-Saharan Africa. Institute of Development Studies, University of Sussex, UK.

Benson, I. and Lloyd, J. 1983. New Technology and Industrial Change. London: Kogan Page.

Berger, S. and Dore, R. (eds) 1996. National Diversity and Global Capitalism. Ithaca, NY: Cornell University Press.

Bergsten, F. 1996. Managing the world economy of the future. In P.B. Kenen (ed.), Managing the World Economy. Washington, DC: Institute for International Economics.

Biggart, N. W. and Hamilton, G. G. 1992. 'On the limits of a firm-based theory to explain business networks'. In N. Nohria and R.G. Eccles (eds), Networks and Organizations. Boston: Harvard Business School Press, 471–90.

Birkinshaw, J.M. and Hood, N. 1998. 'Multinational subsidiary evolution'. Academy of Management Review, 23: 773–95.

Bluestone, B. and Harrison, B. 1982. The Deindustrialization of America. New York: Basic Books.

Bluestone, B. and Harrison, B. 2000. Growing Prosperity. Boston: Century Foundation.

Boltanski, L. and Chiapello, E. 1999. Le Nouvel esprit du capitalisme. Paris: Gallimard.

Bourdieu, P. and Wacquant, L. 2001. 'NewLiberalSpeak: notes on the new planetary vulgate'. Radical Philosophy, 105: 2–5.

Bowker, G. and Star, S.L. 1999. Sorting Things Out. Cambridge, MA: MIT Press.

Boyer, R. and Hollingsworth, J.R. (eds) 1999. Contemporary Capitalism. Cambridge: Cambridge University Press.

Boyne, R. 2001. Subject, Society and Culture. London: Sage.

Braczyk, H.-J., Cooke, P. and Heidenreich, M. (eds) 1998. Regional Innovation Systems. London: UCL Press.

Brahm, R. 1994. 'Commentary: global-local tensions', by P. Dicken. Advances in Strategic Management 10B. Greenwich, CT: JAI Press, 249–54.

Branigan, T. 2001. 'Lifestyle shattered by doubts'. the Guardian 20 October.

Brenner, N. 1998a. 'Global cities, glocal states'. Review of International Political Economy, 5: 1–37.

Brenner, N. 1998b. 'Globalization as reterritorialisation'. Urban Studies, 36: 431–52.

Brenner, N. 2001. Entrepreneurial Cities, Glocalizing States and the New Politics of Scale. Working Paper 76a/76b, Center for European Studies, Harvard University.

Brenner, N. 2003. New State Spaces. Unpublished book manuscript.

Brenner, N. and Theodore, N. (eds) 2002. Spaces of Neoliberalism. Oxford: Blackwell.

Brenner, N., Jessop, B., Jones, M. and MacLeod, G. (eds) 2003. State/Space. Cambridge, MA: Blackwell.

Bridges Weekly Trade News Digest. 2001. 'Implementation: developing countries hit road block'. 20 March.

British Audit Commission. 1998. A Stitch in Time. London: Audit Commission.

Britton, J.N.H. 1999. 'Does nationality still matter?' In T.J. Barnes and M.S. Gertler (eds), The New Industrial Geography. London: Routledge, 238–64.

Bromell, N. 2000. Tomorrow Never Knows. Chicago: University of Chicago Press.

Brown, J.M. 2001. 'Why Irish eyes are still smiling over IT sector'. Financial Times, 7 May.

Brown, B., Green, N. and Harper, R. (eds) 2002. Wireless World. London: Springer Verlag.

Brown J. S. and Duguid, P. 1996. 'Organizational learning and communities-of-practice'. In M. Cohen and L. Sproull (eds), Organizational Learning. London, Sage: 58–82.

Brown, J.S. and Duguid, P. 1998. 'Organizing knowledge'. California Management Review, 40: 90–111.

Brown, J.S. and Duguid, P. 2000. The Social Life of Information. Boston, MA: Harvard Business School Press.

Brown, R. and Raines, P. 2000. 'The changing nature of foreign investment policy in Europe'. In J.H. Dunning (ed.), Regions, Globalization and the Knowledge-Based Economy. Oxford: Oxford University Press, 435–58.

Buchanan, K. 1970. The Transformation of the Chinese Earth. London: Bell.

Buchanan, R.O. 1935. The Pastoral Industries of New Zealand. London: George Philip.

Buckley, P.J. and Casson, M. 1976. The Future of the Multinational Enterprise. London: Macmillan.

Bunnell, T.G. and Coe, N.M. 2001. 'Spaces and scales of innovation'. Progress in Human Geography, 25: 569–89.

Burgess, K. 2002. 'Unloved and unwanted – so buy it'. Financial Times, 23 March.

Cairncross, A. 1994. 'Economic policy and performance, 1964–1990'. In R. Floud and D. McCloskey (eds), The Economic History of Britain Since 1700. Volume 3. Cambridge: Cambridge University Press, 67–94.

Callon, M. 1999. 'Le réseau comme forme émergente et comme modalité de coordination'. In M. Callon, P. Cohendet, N. Curien, J.-M. Dalle, F. Eymard-Duvernay, D. Foray and E. Schenk, Réseau et coordination. Paris: Economica.

Cammack, P. 2002. 'Attacking the poor'. New Left Review, 13: 125–34.

Cane, A. and Nicholson, M. 2001. 'Thousands of job losses feared at Motorola'. Financial Times, 19 April.

Cantwell, J. 1999. 'Innovation as the principal source of growth in the global economy'. In D. Archibugi, J. Howells and J. Michie (eds), Innovation Policy in a Global Economy. Cambridge: Cambridge University Press, 225–41.

Cantwell, J. and Janne, O. 1999. 'Technological globalisation and innovative centres'. Research Policy, 28: 119–44.

Catán, T. 2002. 'Argentines snowed under by paper IOUs'. Financial Times, 11 April.

Cerny, P. 1995. 'Globalization and the changing logic of collective action'. International Organization, 49: 595–625.

CGEA, 2001. Website: www.cgea.com.

Chapman, K. 1992. Review of Global Shift. Scottish Geographical Magazine, 108: 134.

Christopherson, S. 1999. 'Rules as resources'. In T.J. Barnes and M.S. Gertler (eds), The New Industrial Geography. London: Routledge, 155–75.

Christopherson, S. 2002. 'Why do national labor market practices continue to diverge in the global economy?' Economic Geography, 78: 1–20.

Clancy, M. 1998. 'Commodity chains, services and development'. Review of International Political Economy, 5: 122–48.

Clark, A. 2001. Mindware. New York: Oxford University Press.

Clark, G.L., Feldman, M.A. and Gertler, M.S. (eds) 2000. The Oxford Handbook of Economic Geography. Oxford: Oxford University Press.

Clark, T. and Fincham, R. (eds) 2001. Critical Consulting. Oxford: Blackwell.

Clippinger, J.H. 1999. The Biology of Business. San Francisco, CA: Jossey Bass.

Clough, P.T. 2000. Autoaffection. Minneapolis, MN: University of Minnesota Press.

Coe, N. and Kelly, P. F. 2000. 'Connections and constructions in the local labour market'. Area, 32: 413–22.

Coggan, P. 1993. 'Emerging markets investment fashion of the 90s supplement on Global custody'. Financial Times, 7 December.

Conroy, M. 1996. Public Address: The Dark Decade. Department of Geography, Pennsylvania State University, USA.

Conroy, M. and Glasmeier, A. 1994. 'Industrial strategies, the newly industrializing economies, and new international trade theory in Latin America'. Environment and Planning, A, 27: 1–10.

Conway, D. and Cohen, J. 1998. 'Consequences of migration and remittances for Mexican transnational communities'. Economic Geography, 74: 26–44.

Cooke, P. and Morgan, K. 1998. The Associational Economy. Oxford: Oxford University Press.

Corbridge, S. 1993. 'Marxisms, modernities and moralities'. Environment and Planning D, 11: 449–72.

Cox, K.R. (ed.) 1997. Spaces of Globalization. New York: Guilford.

Cox, K.R. 1999. 'Review of Global Shift'. Progress in Human Geography, 23: 474–5.

Crary, J. 1999. Suspensions of Perception. Cambridge, MA: MIT Press.

Crewe, L. and Lowe, M. 1996. 'United colours? Globalization and localization tendencies in fashion retailing'. In N. Wrigley and M. Lowe (eds), Retailing, Consumption and Capital. London: Longman, 271–83.

Dagognet, F. 1992. Etienne-Jules Marey. New York: Zone Books.

Dahl, M.S. 2002. 'Embedded Knowledge Flows Through Labour Market Mobility in Regional Clusters in Denmark'. Paper for the DRUID Summer Conference on 'Industrial Dynamics of the New and Old Economy – who is embracing whom?' Copenhagen/Elsinore, 6–8 June.

Damasio, A. 1999. The Feeling of What Happens. New York: Vintage.

Davenport, T.H. and Beck, J.C. 2001. The Attention Economy. Boston, MA: Harvard Business School Press.

de Nora, T. 2000. Music in Everyday Life. Cambridge: Cambridge University Press.

Deleuze, G. 1993. The Fold: Leibniz and the Baroque. Minneapolis, MN: University of Minnesota Press.

Deleuze, G. and Parnet, C. 1987. Dialogues. London: Athlone.

Deloitte Research. 2001. Global Manufacturing Trends of US Manufacturers. New York: Deloitte Consulting and Deloitte & Touche.

Desai, R. 1994. 'Second-hand dealers in ideas'. New Left Review, 203: 27–64.

Dezalay, Y. and Garth, B.G. 2002. The Internationalization of Palace Wars. Chicago: University of Chicago Press.

Dicken, P. 1971. 'Some aspects of the decision-making behavior of business organizations'. Economic Geography, 47: 426–37.

Dicken, P. 1976. 'The multiplant business enterprise and geographical space'. Regional Studies, 10: 401–12.

Dicken, P. 1977. 'A note on location theory and the large business enterprise'. Area, 9: 138–43.

Dicken, P. 1980. 'Foreign direct investment in European manufacturing industry'. Geoforum, 11: 289–313.

Dicken, P. 1982. 'Recent trends in international direct investment'. In B.T. Robson and J. Rees (eds), Geographical Agenda for a Changing World. London: SSRC.

Dicken, P. 1983. 'Japanese manufacturing investment in the United Kingdom'. Area, 15: 273–84.

Dicken, P. 1986a. Global Shift: Industrial Change in a Turbulent World. London: Harper & Row.

Dicken, P. 1986b. 'Multinational enterprises and the local economy'. Area, 18: 215–21.

Dicken, P. 1987. 'Japanese penetration of the European automobile industry'. Tijdschrift voor Economische en Sociale Geografie, 78: 59–72.

Dicken, P. 1988. 'The changing geography of Japanese foreign direct investment in manufac-turing industry'. Environment and Planning A, 20: 633–53.

Dicken, P. 1990a. 'Transnational corporations and the spatial organization of production'. In A. Shachar and S. Oberg (eds), The World Economy and the Spatial Organization of Power. Aldershot: Avebury, 31–55.

Dicken, P. 1990b. 'The geography of enterprise'. In M. De Smidt and E. Wever (eds), The Corporate Firm in a Changing World Economy. London: Routledge, 234–44.

Dicken, P. 1990c. 'Seducing foreign investors'. In M. Hebbert and J.C. Hansen (eds), Unfamiliar Territory, Aldershot: Gower, 162–86.

Dicken, P. 1992a. Global Shift: The Internationalization of Economic Activity. Second Edition, London: Paul Chapman.

Dicken, P. 1992b. 'International production in a volatile regulatory environment'. Geoforum, 23: 303–16.

Dicken, P. 1992c. 'Europe 1992 and strategic change in the international automobile industry'. Environment and Planning A, 24: 11–32.

Dicken, P. 1994. 'Global-local tensions'. Economic Geography, 70: 101–28. Reprinted in Advances in Strategic Management, 10B. Greenwich, CT: JAI Press, 217–47.

Dicken, P. 1995. 'How the world works'. Review of International Political Economy, 2: 197–204.

Dicken, P. 1997. 'Transnational corporations and nation-states'. International Social Science Journal, 49: 77–90.

Dicken, P. 1998a. Global Shift: Transforming the World Economy. Third Edition, London: Paul Chapman.

Dicken, P. 1998b. 'Globalization: an economic-geographical perspective'. In W.E. Halal and K.R. Taylor (eds), Twenty-First Century Economics. New York: St. Martin's Press, 31–51.

Dicken, P. 2000. 'Places and flows'. In G.L. Clark, M.A. Feldman and M.S. Gertler (eds), The Oxford Handbook of Economic Geography. Oxford: Oxford University Press, 275–91.

Dicken, P. 2002. 'Global Manchester'. In J. Peck and K. Ward (eds), City of Revolution: Restructuring Manchester. Manchester: Manchester University Press, 18–33.

Dicken, P. 2003. Global Shift: Reshaping the Global Economic Map in the 21ˢᵗ Century. Fourth Edition, London: Sage.

Dicken, P. and Hassler, M. 2000. 'Organizing the Indonesian clothing industry in the global economy'. Environment and Planning A, 32: 263–80.

Dicken, P. and Kirkpatrick, C. 1991. 'Services-led development in ASEAN'. The Pacific Review, 4: 174–84.

Dicken, P. and Lloyd, P.E. 1976. 'Geographical perspectives on United States investment in the United Kingdom'. Environment and Planning A, 8: 685–705.

Dicken, P. and Lloyd, P.E. 1980. 'Patterns and processes of change in the spatial distribution of foreign-controlled manufacturing employment in the United Kingdom, 1963–1975'. Environment and Planning A, 12: 1405–26.

Dicken, P. and Lloyd, P.E. 1990. Location in Space. Third Edition, New York: Harper & Row.

Dicken, P. and Malmberg, A. 2001. 'Firms in territories'. Economic Geography, 77: 345–63.

Dicken, P. and Miyamachi, Y. 1998. 'From noodles to satellites': the changing geography of the Japanese sogo shosha'. Transactions of the Institute of British Geographers, 23: 55–78.

Dicken, P. and Quevit, M. (eds) 1994. Transnational Corporations and European Regional Restructuring. Utrecht: Royal Dutch Geographical Society.

Dicken, P. and Thrift, N. 1992. 'The organization of production and the production of organization'. Transactions of the Institute of British Geographers, 17, 279–91.

Dicken, P. and Tickell, A.T. 1992. 'Competitors or collaborators?' Regional Studies, 26: 99–106.

Dicken, P. and Yeung, H.W.C. 1999. 'Investing in the future'. In K. Olds, P. Dicken, P.F. Kelly, L. Kong and H.W.C. Yeung (eds), Globalization and the Asia-Pacific. London: Routledge, 107–28.

Dicken, P., Forsgren, M. and Malmberg, A. 1994. 'The local embeddedness of transnational corporations'. In A. Amin and N. Thrift (eds), Globalization, Institutions, and Regional Development in Europe. Oxford: Oxford University Press, 23–45.

Dicken, P., Peck, J., and Tickell, A. 1997a. 'Unpacking the global'. In R. Lee and J. Wills (eds), Geographies of Economies. London: Arnold, 158–66.

Dicken, P., Tickell, A. and Yeung, H.W.C. 1997b. 'Putting Japanese investment in Europe in its place'. Area, 29: 200–12.

Dicken, P., Kelly, P.F., Olds, K. and Yeung, H.W.C. 2001. 'Chains and networks, territories and scales'. Global Networks, 1: 89–112.

Dolphin, R. 1995. 'Zen and the art of mall raising'. Vancouver, March, 36–47.

Doremus, P. N., Keller, W. W., Pauly, L. W. and Reich, S. 1998. The Myth of the Global Corporation. Princeton, NJ: Princeton University Press.

Doyle, R. 1997. Beyond Living. Stanford, CA: Stanford University Press.

Duffy, F. 1997. The New Office. London: Octopus.

Dunford, M., Hudson R. and Smith, A. 2001. 'Restructuring the European Clothing Sector'. Sussex European Institute, University of Sussex, UK, mimeo.

Dunning, J.H. 1977. 'Trade, location of economic activity and the MNE'. In B. Ohlin, P.O. Hesselborn and P.M. Wijkman (eds), The International Allocation of Economic Activity. London: Macmillan, 395–418.

Dunning, J.H. 1979. 'Explaining changing patterns of international production'. Oxford Bulletin of Economics and Statistics, 41: 269–96.

Dunning, J.H. 1993. Multinational Enterprises and the Global Economy. Reading, MA: Addison Wesley.

Dunning, J.H. 1998. 'Location and the multinational enterprise'. Journal of International Business Studies, 29: 45–66.

Dunning, J.H. (ed.) 2000. Regions, Globalization and the Knowledge-based Economy. Oxford: Oxford University Press.

Edquist, C. (ed.) 1997. Systems of Innovation. London: Pinter.

Egelhoff, W.G. 1982. 'Strategy and structure in multinational corporations'. Administrative Science Quarterly, 27: 435–58.

Eichengreen, B. 1998. Globalizing Capital. Princeton, NJ: Princeton University Press.

Eichengreen, B. and Kenen, P. 1996. 'Managing the world economy under the Bretton Woods system'. In P.B. Kenen (ed.), Managing the World Economy. Washington, DC: Institute for International Economics.

Eisenschitz, A. and Gough, J. 1993. The Politics of Local Economic Development. New York: Macmillan.

Eliasson, G. 2000. 'Industrial policy, competence blocks and the role of science in economic development'. Journal of Evolutionary Economics, 10: 217–41.

Enright, M.J. 1998. 'Regional clusters and firm strategy'. In A.D. Chandler, P. Hagström and Ö. Sölvell (eds), The Dynamic Firm. Oxford: Oxford University Press, 315–42.

Esty, D.C. 2000. WTO Legitimacy Beyond the Club Model: A Comment on Keohane and Nye. Conference on 'Efficiency, Equity, and Legitimacy': The Multilateral Trading System at the Millennium, Harvard University, 1–2 June.

European Commission. 1999. 'The EU Eco-Industry's Export Potential. Final Report to DGXI of the European Commission'. September.

Evans, P. 2000. 'Fighting marginalization with transnational networks'. Contemporary Sociology, 29: 230–41.

Feigenbaum, H., Henig, J. and Hamnett, C. 1998. Shrinking the State. Cambridge: Cambridge University Press.

Ferner, A., Quintanilla, J. and Varul, M. Z. 2001. 'Country-of-origin effects, host-country effects, and the management of HR in multinationals'. Journal of World Business, 36: 107–27.

Financial Times. 1995. 'Markets catch a chill'. Financial Times, 11 January.

Financial Times. 2001. 'Vivendi sells stake in its environment division'. Financial Times, 5 December:

Financial Times. 2002a. 'German utility makes approach to UK power group'. Financial Times, 18 February:

Financial Times. 2002b. 'Vivendi expected to sell Environment stake'. Financial Times, 20 April.

Finger, J.M. and Schuler, P. 1999. Implementation of the Uruguay Round Commitments. Paper presented at the Conference on WTO Negotiations, Geneva.

Fischer, R. 2001. A History of Writing. London: Reaktion.

Florida, R. 1997. 'The globalization of R&D'. Research Policy, 26: 85–103.

Fransman, M. 1994. 'Information, knowledge, vision and theories of the firm'. Industrial and Corporate Change, 3: 713–27.

Freeman, C. 1997. 'The national system of innovation in historical perspective'. In D. Archibugi and J. Michie (eds), Technology, Globalisation and Economic Performance. Cambridge: Cambridge University Press, 24–49.

Fröbel, F., Heinrichs, J. and Kreye, O. 1980. The New International Division of Labour. Cambridge: Cambridge University Press.

Fuellhart, K. 1999. 'Localization and the use of information sources'. European Urban and Regional Studies, 6: 39–58.

Fujita, M., Krugman, P. and Venables, A. 1999. The Spatial Economy. Cambridge, MA: MIT Press.

Furusten, S. 1999. Popular Management Books. London: Routledge.

Galbraith, J.K. 1994. A Short History of Financial Euphoria. Harmondsworth: Penguin.

Galison, P. and Thompson, E. (eds) 2000. The Architecture of Science. Cambridge, MA: MIT Press.

Garofoli, G. 2002. 'Local development in Europe'. European Urban and Regional Studies, 9: 225–40.

Gerlach, M. L. and Lincoln, J. R. 1992. 'The organization of business networks in the United States and Japan'. In N. Nohria and R.G. Eccles (eds), Networks and Organizations. Boston, MA: Harvard Business School Press, 491–520.

Geroski, P. and Gugler, K.P. 2001. Corporate Growth Convergence in Europe. Discussion Paper 2838, London: Centre for Economic Performance.

Gertler, M.S. 1988. 'The limits to flexibility'. Transactions of the Institute of British Geographers, 13, 419–32.

Gertler, M.S, 1995. 'Being there: proximity, organization and culture in the development and adoption of advanced manufacturing technologies'. Economic Geography, 70: 1–26.

Gertler, M.S. 1996. 'Worlds apart: the changing market geography of the German machinery industry'. Small Business Economics, 8: 87–106.

Gertler, M.S. 1999. 'The production of industrial processes'. In T.J. Barnes and M.S. Gertler (eds), The New Industrial Geography: Regions, Regulation and Institutions. London: Routledge, 225–37.

Gertler, M.S. 2001. 'Best practice? Geography, learning and the institutional limits to strong convergence'. Journal of Economic Geography, 1: 5–26.

Gertler, M.S. 2002a. 'Technology, culture and social learning'. In M.S. Gertler and D.A. Wolfe (eds), Innovation and Social Learning. Basingstoke: Palgrave/Macmillan, 111–34.

Gertler, M.S. 2002b. 'Tacit knowledge and the economic geography of context'. Journal of Economic Geography, 2 (forthcoming).

Gertler, M.S. 2002c. Tacit Knowledge, Path-Dependency and Local Growth Trajectories. Paper presented at the workshop on 'Rethinking regional innovation and change: Path-dependency or regional breakthrough', Stuttgart, Germany, 29 February–1 March.

Gertler, M.S. and DiGiovanna, S. 1997. 'In search of the new social economy'. Environment and Planning A, 29: 1585–602.

Gertler, M.S., Wolfe, D.A. and Garkut, D. 2000. 'No place like home?'. Review of International Political Economy, 7: 1–31.

Gettleman, J. 1996. 'Global find managers make tracks for Africa'. Financial Times, 1 February.

Giddens, A. 1998. The Third Way. Cambridge: Polity Press.

Gilmore, R. W. 1998. 'Globalization and US prison growth'. Race and Class, 40: 171–88.

Girard, L. and Stark, D. 2002. 'Heterarchy in the media industry'. Environment and Planning A, 33 (forthcoming).

Glaeser, E.L., Kallal, H.D., Scheinkman, J. and Shleifer, A. 1992. 'Growth in cities'. Journal of Political Economy, 100: 1126–52.

Glasmeier, A.K. 2001. Manufacturing Time. New York: Guilford Press.

Glasmeier, A.K., Thompson, J. and Kays, A. 1993. The geography of trade policy. Transactions of the Institute of British Geographers, 18: 19–35.

Gnyawali, D.R. and Madhavan, R. 2001. 'Cooperative networks and competitive dynamics'. Academy of Management Review, 26: 431–45.

Gonzales, J. 1998. Philippine Labour Migration. Singapore: Institute of South East Asian Studies.

Gourevitch, P. 1990. Politics in Hard Times. Ithaca, NY: Cornell University Press.

Grabher, G. (ed.) 1993. The Embedded Firm. London: Routledge.

Grabher, G. 2001. 'Ecologies of creativity'. Environment and Planning A, 33: 351–74.

Granovetter, M. 1985. 'Economic action and social structure'. American Journal of Sociology, 91: 481–510.

Granstrand, O. 1999. 'Internationalization of corporate R&D'. Research Policy, 28: 275–302.

Grant, R.M. 1996. Prospering in dynamically-competitive environments. Organization Science, 7: 375–87.

Griffith, C. 2000. A Citizen's Handbook to the Office of the United States Trade Representative. Washington, DC: Consumer Choice Council.

Guerrera, F. 2001. 'Brussels' face set against "national champions"'. Financial Times, 3 October.

Guerrera, F. and Mallet, V. 2001. 'Brussels blocks Schneider deal with Legrand'. Financial Times, 11 October.

Gulati, R. and Gargiulo, M. 1999. 'Where do interorganizational networks come from?' American Journal of Sociology, 104: 1439–93.

Gupta, A.K. and Govindarajan, V. 1991. 'Knowledge flows and the structure of control within multinational corporations'. Academy of Management Review, 16: 768–92.

Gupta, A.K. and Govindarajan, V. 2000. 'Knowledge flows within multinational corporations'. Strategic Management Journal, 21: 473–96.

Hadjimichalis, C. 1998. Small and Medium Enterprises in Greece. University of Thessaloniki, Greece, mimeo.

Hadjimichalis, C. and Papamichos, N. 1990. 'Local development in southern Europe'. Antipode, 22: 181–210.

Hägerstrand, T. 1970. 'What about people in regional science?' Regional Science Association Papers, XXIV: 7–21.

Haggard, S. 1990. Pathways from the Periphery. Ithaca, NY: Cornell University Press.

Haggett, P. 1965. Locational Analysis in Human Geography. London: Edward Arnold.

Håkansson, H. 1987. Corporate Technological Behaviour. London: Routledge.

Hall, P. and Soskice, D. (eds) 2001. Varieties of Capitalism. Oxford: Oxford University Press.

Hall, T. and Hubbard, P. (eds) 1998. The Entrepreneurial City. London: John Wiley.

Hamilton, F.E.I. (ed.) 1974. Spatial Perspectives on Industrial Organization and Decision-Making. London: John Wiley.

Hamilton, G.G. 1996. 'Overseas Chinese capitalism'. In W.-M. Tu (ed.), Confucian Traditions in East Asian Modernity. Cambridge, MA: Harvard University Press, 328–42.

Hamilton, G.G. (ed.) 1999. Cosmopolitan Capitalists. Seattle, WA: University of Washington Press.

Hamilton, G.G. 2000. 'Reciprocity and control: the organization of Chinese family-owned conglomerates'. In H.W.C. Yeung and K. Olds (eds), The Globalisation of Chinese Business Firms. London: Macmillan, 55–74.

Hamilton, G.G. and Feenstra, R.C. 1998. 'Varieties of hierarchies and markets', In G. Dosi, D.J. Teece and J. Chytry (eds), Technology, Organization and Competitiveness. Oxford: Oxford University Press, 105–46.

Hannerz, U. 1996. Transnational Connections. London: Routledge.

Hansen, N. 2000. Embodying Technesis. Ann Arbor: University of Michigan Press.

Harding, A. 1997. 'Urban regimes in a Europe of the cities?' European Urban and Regional Studies, 4: 291–314.

Hardt, M. and Negri, A. 2001. Empire. Cambridge, MA: Harvard University Press.

Hargadon, A. 1998.' Firms as knowledge brokers'. California Management Review, 40: 209–28.

Harloe, M. 2001. 'Social justice and the city: the new liberal formulation'. International Journal of Urban and Regional Research, 25: 889–97.

Harris, S. J. 1998. 'Long-distance corporations, big sciences, and the geography of knowledge'. Configurations, 6: 269–304.

Hartwick, E. 1998. 'Geographies of consumption'. Environment and Planning D, 16: 423–37.

Harvey, D. 1982. The Limits to Capital. Oxford: Basil Blackwell.

Harvey, D. 1989a. 'From managerialism to entrepreneurialism'. Geografiska Annaler B, 71: 3–18.

Harvey, D. 1989b. The Urban Experience. Baltimore: The Johns Hopkins University Press.

Harvey, D. 2001 [1975]. 'The geography of capitalist accumulation'. In D. Harvey, Spaces of Capital. Edinburgh: Edinburgh University Press, 237–66.

Hatch, M.-J. 1999. 'Exploring the empty spaces of organizing'. Organization Studies, 20: 75–100.

Hayek, F.A. 1944. The Road to Serfdom. London: Routledge.

Hayes, J. and Allison, C. W. 1998. 'Cognitive style and the theory and practice of individual and collective learning in organizations'. Human Relations, 51: 847–72.

Held, D., McGrew, A., Goldblatt, D. and Perraton, J. 1999. Global Transformations. Cambridge: Polity Press.

Henderson, J., Dicken, P., Hess, M., Coe, N. and Yeung, H.W.C. 2002. 'Global production networks and the analysis of economic development'. Review of International Political Economy, 9 (forthcoming).

Herrigel, G. 2000. 'Large firms and industrial districts in Europe'. In J.H. Dunning (ed.), Regions, Globalization and the Knowledge-Based Economy. Oxford: Oxford University Press, 286–302.

Hill, C. 1975. The World Turned Upside Down. Harmondsworth: Penguin.

Hirst, P. and Thompson, G. 1996. Globalization in Question. Cambridge: Polity Press.

Hobart, M.E. and Schiffman, Z. 1998. Information Ages. Baltimore: The Johns Hopkins University Press.

Hofstede, G. 1980. Culture's Consequences. London: Sage

Hofstede, G. 1983. 'The cultural relativity of organizational practices and theories'. Journal of International Business Studies, 14: 75–89.

Hollingsworth, J.R. 1997. 'Continuities and changes in social systems of production'. In J.R. Hollingsworth and R. Boyer (eds), Contemporary Capitalism, Cambridge: Cambridge University Press, 265–318.

Hollingsworth, J.R. and Boyer, R. (eds) 1997. Contemporary Capitalism. Cambridge: Cambridge University Press.

Holm, U. and Pedersen, T. (eds) 2000. The Emergence and Impact of MNC Centres of Excellence. London: Macmillan.

Hsing, Y.-T. 1998. Making Capitalism in China. New York: Oxford University Press.

Hsu, J.-Y., and Saxenian, A. 2000. 'The limits of guanxi capitalism: transnational collaboration between Taiwan and the USA'. Environment and Planning A, 32: 1991–2005.

Hu, Y.-S. 1992. 'Global firms are national firms with international operations'. California Management Review, 34: 107–26.

Hudson, R. 1999. 'The new economy of the New Europe'. In R. Hudson and A. Williams (eds), Divided Europe. London: Sage, 29–62.

Hudson, R. 2000. 'One Europe or many?' Transactions of the Institute of British Geographers, 25: 409–26.

Hudson, R. 2001a. Producing Places. New York: Guilford.

Hudson, R. 2001b. 'Regional Development, Flows of Value and Governance Processes in an Enlarged Europe'. Working Paper 6–01, Regional Economic Performance, Governance and Cohesion in an Enlarged Europe, University of Sussex, UK. Available at: http://www.geog.susssex.ac.uk/research/changing-Europe/wprkpaper.html

Hudson, R. 2002. 'Changing industrial production systems and regional development in the New Europe'. Transactions of the Institute of British Geographers, 27 (forthcoming).

Hudson, R. and Williams, A. 1999. 'Re-shaping Europe'. In R. Hudson and A. Williams (eds), Divided Europe. London: Sage, 1–28.

Hui, W.T. 1997. 'Regionalization, economic restructuring and labour migration in Singapore'. International Migration, 35: 109–28.

Hutchins, E. 1995. Cognition in the Wild. Cambridge, MA: Harvard University Press.

Hutchins, E. 1996. 'Organizing work by adaptation'. In M. Cohen and L. Sproull (eds), Organizational Learning. London: Sage, 20–57.

Hymer, S.H. [1960] 1976. The International Operations of National Firms. Cambridge, MA: MIT Press.

Ichijo, K., von Krogh, G. and Nonaka, I. 1998. 'Knowledge enablers'. In G. von Krogh, J. Roos and D. Kleine (eds), Knowing in Firms. London: Sage, 173–203.

IMF [International Monetary Fund]. 2001. Transparency at the International Monetary Fund. Washington, DC: IMF.

Itzigsohn, J., Dore, C., Hernandez, E. and Vazquez, O. 1999. 'Mapping Dominican transnationalism'. Ethnic and Racial Studies, 22: 316–39.

Jackson, J. 1996. 'Managing the trading system'. In P.B. Kenen (ed.), Managing the World Economy. Washington, DC: Institute for International Economics.

Jackson, J. 2001. 'The role and effectiveness of the WTO dispute settlement mechanism'. In S.M. Collins and D. Rodrik (eds), Brookings Trade Forum 2000. Washington, DC: Brookings Institution.

Jessop, B. 1990. State Theory. Cambridge: Polity Press.

Jessop, B. 1994. 'Post-Fordism and the state'. In A. Amin (ed.), Post-Fordism. Cambridge, MA: Blackwell, 251–79.

Jessop, B. 1997. 'Capitalism and its future'. Review of International Political Economy, 4: 561–81

Jessop, B. 1999. 'Narrating the future of the national economy and the national state'. In G. Steinmetz (ed.), State/Culture. Ithaca, NY, Cornell University Press, 378–405.

Jessop, B., Bonnett, K., Bromley, S. and Ling, T. 1988. Thatcherism. Cambridge: Polity.

Johnson, C. 1999. 'Ambient technologies, uncanny signs'. Oxford Literary Review, 21: 117–34.

Johnson, S. 2001. Emergence. London: Faber and Faber.

Jones, M. 1997. 'Spatial selectivity of the state?' Environment and Planning A, 29: 831–64.

Katz, J. 1999. How Emotions Work. Chicago: University of Chicago Press.

Katz, J. and Ackhus, M. (eds) 2002. Perpetual Contact. Cambridge: Cambridge University Press.

Kay, L. 2000. Who Wrote the Book of Life? Stanford, CA: Stanford University Press.

Keating, M. 1997. 'The invention of regions'. Environment and Planning C, 15: 383–98.

Kelly, P.F. 1999. 'The geographies and politics of globalization'. Progress in Human Geography, 23: 379–400.

Keohane, R. O. and Nye, J. Jr. 2000. 'The Club Model of Multilateral Cooperation in the World Trade Organization', Conference on 'Efficiency, Equity, and Legitimacy: The Multilateral Trading System at the Millennium, Harvard University, 1–2 June.

Kindleberger, C. 1969, American Business Abroad. New Haven, CT: Yale University Press.

Kittler, F. 1997. Literature, Media, Information Systems. Amsterdam: G and B Arts.

Kobrin, S.J. 2001. 'Sovereignty@bay'. In A.M. Rugman and T.L. Brewer (eds), Oxford Handbook of International Business. Oxford: Oxford University Press, 181–205.

Kogut, B. and Singh, H. 1988. 'The effect of national culture on the choice of entry mode'. Journal of International Business Studies, 19: 411–32.

Krumme, G. 1987. 'Review of Global Shift'. Environment and Planning A, 19: 132–3.

Kuemmerle, W. 1999. 'Foreign direct investment in industrial research in the pharmaceutical and electronics industries'. Research Policy, 28: 179–93.

Lakoff, G. and Johnson, M. 1999. Philosophy in the Flesh. New York: Basic Books.

Lam, A. 1998. 'Tacit Knowledge, Organisational Learning and Innovation'. DRUID Working Paper No. 98–22, Aalborg University, Denmark.

Lam, A. 2000. 'Tacit knowledge, organizational learning and societal institutions'. Organization Studies, 21: 487–513.

Lam, A. and Lundvall, B-Å. 2000. 'Innovation Policy and Knowledge Management in the Learning Economy'. Paper presented at the OECD High Level Forum on Knowledge Management, Ottawa, 21 September.

Landolt, P., Autlet, L., and Baires, S. 1999. 'From Hermano Lejano to Hermano Mayor'. Ethnic and Racial Studies, 22: 290–315.

Lane, C. 1997. 'The social regulation of inter-firm relations in Britain and Germany'. Cambridge Journal of Economics, 21: 197–215.

Larner, W. 2000. 'Theorising neo-liberalism'. Studies in Political Economy, 63: 5–26.

Larsson, S. 1998. 'Lokal förankring och global räckvidd'. En studie av teknikutveckling i svensk maskinindustri, Geografiska regionstudier nr. 35, Uppsala: Kulturgeografiska institutionen, Uppsala universitet.

Larsson, S. and Lundmark, M. 1991. 'Kista – företag i nätverk eller statusadress?' En studie av Kistaföretagens länkningar. Forskningsrapport nr. 100, Uppsala: Kulturgeografiska institutionen, Uppsala universitet.

Lash, S. 2002. Critique of Information. London, Sage.

Latour, B. 1986. Visualisation and cognition. Knowledge and Society, 6: 1–40.

Latour, B. and Woolgar, S. 1979. Laboratory Life. Beverly Hills, CA: Sage.

Lawson, C. 1999. Towards a competence theory of the region. Cambridge Journal of Economics, 23:151–66.

Lazonick, W. 2000. From Innovative Enterprise to National Institutions. Working Paper, Corporate Governance, Innovation and Economic Performance in the EU, Fontainebleau: INSEAD.

Le Doux, J. 1998. The Emotional Brain. London: Weidenfeld and Nicolson.

Leamer, E. and Storper, M. 2001. The Economic Geography of the Internet Age. NBER Working Paper W8450, Cambridge, MA: National Bureau of Economic Research.

Lee, R. 1989. Social relations and the geography of material life. In D. Gregory and R. Walford (eds), Horizons in Human Geography. London: Macmillan, 152–69.

Lee, R. 1990. Making Europe. In M. Chisholm and D.M. Smith (eds), Shared Space Divided Space. London: Unwin Hyman, 235–59.

Lee, R. 1996. Moral money? Environment and Planning A, 28: 1377–94.

Lee, R. 1999. Local money. In R. Martin (ed.), Money and the Space Economy. Chichester: John Wiley, 207–24.

Lee, R. 2002. 'Nice maps, shame about the theory': Thinking geographically about the economic. Progress in Human Geography, 25: 333–55.

Lee, R. and Wills, J. (eds) 1997. Geographies of Economies. London: Edward Arnold.

Lefebvre, H. 1978. De l'état. Volume 4. Paris: Union Générale d'Éditions.

Leitner, H. and Sheppard, E. 1997. Economic uncertainty, inter-urban competition and the efficacy of entrepreneurialism. In T. Hall and P. Hubbard (eds), The Entrepreneurial City. Chichester: John Wiley, 285–308.

Lessig, L. 2002. The Fate of Ideas. New York: Random House.

Levitt, P. 2001. The Transnational Villagers. Berkeley, CA: University of California Press.

Levitt, T. 1983. The globalization of markets. Harvard Business Review, May-June: 92–102.

Ley, D. forthcoming. Seeking homo economicus: the strange story of Canada's Business Immigration Program. Annals of the American Association of Geographers.

Leys, C. 1994. Confronting the African tragedy. New Left Review, 204: 33– 47.

Leyshon, A. 1994. Review of Global Shift. Progress in Human Geography, 18: 110–11.

Leyshon, A. 1997. Introduction: true stories? In R. Lee and J. Wills (eds), Geographies of Economies. London: Edward Arnold, 133–46.

Lloyd, P.E. and Dicken, P. 1972/1977. Location in Space. First and Second Edition, New York: Harper & Row.

Luce, E. 2001. Hot money flooding into Pakistan. Financial Times, 3 October.

Lundvall, B.-Å. (ed.) 1992. National Systems of Innovation. London: Pinter.

Lundvall, B.-Å. and Johnson, B. 1994. The learning economy. Journal of Industry Studies, 1: 23–42.

Lundvall, B.-Å. and Maskell, P. 2000. Nation states and economic development. In G.L. Clark, M.P. Feldman, and M.S. Gertler (eds), The Oxford Handbook of Economic Geography. Oxford: Oxford University Press, 353–72.

Mackinnon, D., Cumbers, A. and Chapman, K. 2003. Networks, learning and embeddedness amongst SMEs in the Aberdeen oil complex. Entrepreneurship and Regional Development, 15 (forthcoming).

MacLeod, G. 2000. The learning region in an age of austerity. Geoforum, 31: 219–36.

MacLeod, G. and Goodwin, M. 1999. Space, scale and state strategy. Progress in Human Geography, 23: 503–27.

Maddison, A. 2001. The World Economy. Paris: Organisation for Economic Cooperation and Development.

Malmberg, A. and Maskell, P. 2002. The elusive concept of localization economies. Environment and Planning A, 34: 429–49.

Malmberg, A. and Power, D. 2002. On the Role of Global Demand in Local Innovation Processes. Paper presented at the workshop on 'Rethinking regional innovation and change: Path-dependency or regional breakthrough', Stuttgart, Germany, 29 February–1 March.

Malmberg, A. and Sölvell, Ö. 2002. Does foreign ownership matter? In V. Havila, M. Forsgren and H. Håkansson (eds), Critical Perspectives on Internationalisation. Amsterdam: Elsevier, (forthcoming).

Malmberg, A., Sölvell, S. and Zander, I. 1996. Spatial Clustering, local accumulation of knowledge and firm competitiveness. Geografiska Annaler B, 78: 85–97.

Malmberg, A., Malmberg, B. and Lundequist, P. 2000. Agglomeration and firm performance. Environment and Planning A, 32: 305–21.

Mandel-Campbell, A. 2000. Boost as Mexico makes the grade. Financial Times, 27 March.

Mann, M. 1993. Nation-states in Europe and other continents. Proceedings of the American Academy of Arts and Sciences, 122: 155–40.

Mann, M. 2001. Brussels tries again for agreement on cross-border takeover rules. Financial Times, 5 September.

Marino, M. and Boland, J. 1999. An Integrated Approach to Wastewater Treatment. Washington, DC: World Bank.

Markgren, B. 2001. Är närhet en geografisk fråga? Företags affärsverksamhet och geografi – en studie av beroenden mellan företag och lokaliseringens betydelse. Doctoral Thesis No. 85, Företagsekonomiska institutionen, Uppsala universitet.

Markusen, A. 1996. Sticky places in slippery space. Economic Geography, 72: 293–313.

Markusen, A. 1999. Fuzzy concepts, scanty evidence, policy distance. Regional Studies, 33: 869–84.

Marshall, A. 1890. Industrial organization, continued. The concentration of specialized industries in particular localities. In A. Marshall (ed.), Principles of Economics, Book IV, Chapter X. London: Macmillan.

Martin, R. and Sunley, P. 1997. The post-Keynesian state and the space economy. In R. Lee and J. Wills (eds), Geographies of Economies. London: Edward Arnold, 278–289.

Martin, R. and Sunley, P. 2001. Deconstructing Clusters. Paper presented at the RSA Conference on Regionalising the Knowledge Economy, London, 21 November.

Marx, K. 1996 [1867]. Capital. Volume 1, Harmondsworth: Penguin.

Maskell, P. and Malmberg, A. 1999. Localised learning and industrial competitiveness. Cambridge Journal of Economics, 23: 167–86.

Maskell, P., Eskelinen, H., Hannibalsson, I., Malmberg, A. and Vatne, E. 1998. Competitiveness, Localised Learning and Regional Development. London: Routledge.

Massey, D. and Meegan, R. 1979. The Anatomy of Job Loss. London: Methuen.

Maturana, H. and Varela, F. 1980. Autopoesis and Cognition. Dordrecht, Holland: Reidel.

Maurice, M., Sellier, F. and Silvestre, J.-J. 1986. The Social Foundations of Industrial Power. Cambridge, MA: MIT Press.

May, J. and Thrift, N.J. (eds) 2001. Timespace. Geographies of Temporality. London: Routledge.

McConnell, J. and McPherson, A. 1994. The North American Free Trade Area. In R. Gibb and W. Michalak (eds), Continental Trading Blocks. New York: John Wiley, 163–88.

McIlwaine, C. and Willis, K. (eds) 2002. Challenges and Change in Middle America. Harlow: Pearson Education.

McNeil, J. 1992. Keeping Together in Time. New York: Harper & Row.

Micklethwait, J. and Wooldridge, A. 1996. The Witch Doctors. New York: Times Books.

Mitchell, K. 1995. Flexible circulation in the Pacific Rim. Economic Geography, 71: 364–82.

Mitchell, K. 1997. Different diasporas and the hype of hybridity. Environment and Planning D, 15: 533–53.

Mitchell, K. and Olds, K. 2000. Chinese business networks and the globalization of property markets in the Pacific Rim. In H.W.C. Yeung and K. Olds (eds), The Globalisation of Chinese Business Firms. London: Macmillan, 195–219.

Mittelman, J.H. 2000. The Globalization Syndrome. Princeton, NJ: Princeton University Press.

Mohan, G., Brown, E., Milward, B. and Zach-Williams, A.B. 2000. Structural Adjustment. London, Routledge.

Moody, G. 2001. Rebel Code. Cambridge, MA: Perseus Publishing.

Moore, M. 2001. The WTO: Challenges Ahead. Comments of, Director General-the World Trade Organization. Conference on the Role of the WTO in Global Governance. Geneva, Switzerland. May.

Morgan, K. 1997. The learning region. Regional Studies, 31: 491–503.

Morgan, K. 2001. The Exaggerated Death of Geography. Paper presented to The Future of Innovation Studies Conference, The Eindhoven Centre for Innovation Studies, Eindhoven University of Technology, 20–23 September.

Morris, H. 2001. A dynamic approach to the world's grey areas. Financial Times, 19 March.

Mullings, B. 1999. Sides of the same coin: coping and resistance strategies among Jamaican data-entry operators. Annals of the Association of American Geographers, 89: 290–311.

Nachum, L. 2000. Economic geography and the location of TNCs. Journal of International Business Studies, 31: 367–85.

Nardi, B.A. and O'Day, V.L. 2001. Information Ecologies. Cambridge, MA: MIT Press.

Nelson, R.R. (ed.) 1993. National Innovation Systems. New York: Oxford University Press.

Nelson, R.R. 1995. Recent evolutionary theorizing about economic change. Journal of Economic Literature, 33: 48–90.

Nilsson, J.-E., Dicken, P. and Peck, J. (eds) 1996. The Internationalization Process. London: Paul Chapman.

Nitzan, J. 2001. Regimes of differential accumulation. Review of International Political Economy, 8: 226–74.

Nohria, N. and Ghoshal, S. 1997. The Differentiated Network. San Francisco, CA: Jossey-Bass.

Nolan, J.L. 1998. State. Albany, NY: State University of New York Press.

Nonaka, I. 1994. A dynamic theory of organizational knowledge creation. Organization Science, 5: 14–37.

Nonaka, I. and Konno, N. 1998. The concept of 'ba'. California Management Review, 40: 40–54.

Nonaka, I. and Takeuchi, H. 1995. The Knowledge-Creating Company. Oxford: Oxford University Press.

Nooteboom, B. 2000. Learning by Interaction. Rotterdam School of Management, Erasmus University, Rotterdam. mimeo.

Norman, D.A. 1999. The Invisible Computer. Cambridge, MA: MIT Press.

Nye, J. 1996. Comments in response to Fred Bergsten's presentation. In P.B. Kenen (ed.), Managing the World Economy. Washington, DC: Institute for International Economics.

OECD. 1996. The Global Environmental Goods and Services Industry. Paris: Organisation for Economic Cooperation and Development.

OECD. 1998. Internationalisation of Industrial R&D. Paris: Organisation for Economic Cooperation and Development.

OECD. 2000. Knowledge Management in the Learning Society. Paris: Organisation for Economic Cooperation and Development.

Ohmae, K. 1985. Triad Power. New York: The Free Press.

Ohmae, K. 1990. The Borderless World. London: Collins.

Ohmae, K. 1995. The End of the Nation-State. London: HarperCollins.

Oinas, P. 2000. Distance and learning: does proximity matter? In F. Boekma, K. Morgan, S. Bakkers and R. Rutten (eds), Knowledge, Innovation and Economic Growth. Cheltenham: Edward Elgar, 57–72.

Olds, K. 2001. Globalization and Urban Change. Oxford: Oxford University Press.

Olds, K. and Yeung, H.W.C. 1999. (Re)shaping 'Chinese' business networks in a globalising era. Environment and Planning D, 17: 535–55.

Olds, K., Dicken, P., Kelly, P.F., Kong, L. and Yeung, H.W.C. (eds) 1999. Globalization and the Asia-Pacific. London: Routledge.

O'Neill, P. M. 1997. Bringing the qualitative state into economic geography. In R. Lee and J. Wills (eds), Geographies of Economies. London: Edward Arnold, 290–301.

Ong, A. 1999. Flexible Citizenship. Durham, NC: Duke University Press.

Orbinski, J. 2001. Health, equity, and trade. In G. Sampson (ed.), The Role of the World Trade Organization in Global Governance. Geneva: United Nations University, 223–42.

Orr, J. 1996. Talking about Machines. Ithaca, NY: IRL Press.

Ostry, S. and Nelson, R.R. 1995. Techno-Nationalism and Techno-Globalism. Washington, DC: Brookings Institution.

O'Sullivan, M. 2000. Contests for Corporate Control. Oxford: Oxford University Press.

Ó Tuathail, G. 1997. Emerging markets and other simulations. Ecumene, 4: 300–17.

Owen-Smith, J. and Powell, W.W. 2002. Knowledge Networks in the Boston Biotechnology Community. Unpublished working paper, Stanford University.

Painter, J. and Goodwin, M. 1996. Local governance and concrete research. Economy and Society, 24: 334–56.

Palloix, C. 1975. The internationalization of capital and the circuit of social capital. In H. Radice (ed.), International Firms and Modern Imperialism, Harmondsworth: Penguin, 63–88.

Pashler, H.E. 1998. The Psychology of Attention. Cambridge, MA: MIT Press.

Patel, P. and Pavitt, K. 1994. Uneven (and divergent) technological accumulation among advanced countries. Industrial and Corporate Change, 3: 759–87.

Patel, P. and Pavitt, K. 1997. The technological competencies of the world's largest firms. Research Policy, 26: 141–56.

Patel, P. and Vega, M. 1999. Patterns of internationalisation of corporate technology. Research Policy, 28: 145–55.

Pauly, L.W. and Reich, S. 1997. National structures and multinational corporate behaviour, International Organization, 51: 1–30.

Pavitt, K. and Patel P, 1991. Large firms in the production of the world's technology. Journal of International Business Studies, 22: 1–21.

Pavitt, K. and Patel, P. 1999. Global corporations and national systems of innovation. In D. Archibugi, J. Howells and J. Michie (eds), Innovation Policy in a Global Economy. Cambridge: Cambridge University Press, 94–119.

Pearce, R.D. 1999. Decentralized R&D and strategic competitiveness. Research Policy, 28: 157–78.

Pearson, R. 2001. The Red Global de Treuque – Argentina. Notes presented to the Latest Developments in LES and Time Money Conference, Local Economic Policy Unit, South Bank University, London, 4–5 July.

Peck, J. 1996. Work-Place. New York: Guilford.

Peck, J. 1998. Geographies of governance: TEC, and the neoliberalisation of 'local interests'. Space & Polity, 2: 5–31.

Peck, J. 2000. Doing regulation.' In G.L. Clark, M.P. Feldman and M.S. Gertler (eds), The Oxford Handbook of Economic Geography. Oxford: Oxford University Press, 60–80.

Peck, J. and Tickell, A. 1994. Searching for a new institutional fix. In A. Amin (ed.), Post-Fordism. Cambridge, MA: Blackwell, 280–315.

Peck, J. and Tickell, A. 1995. The social regulation of uneven development. Environment and Planning A, 27: 15–40.

Peck, J. and Tickell, A. 2002. Neoliberalizing space. Antipode, 34: 380–404.

Perez, E. 1998. One Year into the Asian Crisis: The Impact on Mexico. UN Department of Economic and Social Affairs, www.un.org/esa/analysis/eperez.pdf, accessed on 25 March 2002.

Pfirrmann, O. 1999. Review of Global Shift. Growth and Change, 30: 155–6.

Phelps, N.A. and Fuller, C. 2000. Multinationals, intracorporate competition, and regional development. Economic Geography, 76: 224–43.

Phelps, N. and MacKinnon, D. 2000. Industrial enclaves or embedded firms? Contemporary Wales, 13: 46–67.

Phelps, N., Lovering, J. and Morgan, K. 1998. Tying the firm to the region or tying the region to the firm? European Urban and Regional Studies, 5: 119–37.

Piore, M.J. and Sabel, C.F. 1984. The Second Industrial Divide. New York: Basic Books.

Polanyi, K. 1944. The Great Transformation. Boston: Beacon Press.

Popper, M. and Lipshitz, R. 1998. Organizational learning mechanisms. Journal of Applied Behavioural science, 34: 161–79.

Porter, M. E. 1990. The Competitive Advantage of Nations. Basingstoke: Macmillan.

Porter, M. E. 1994. The role of location in competition. Journal of the Economics of Business, 1: 35–9.

Porter, M. E. 1998. Clusters and the new economics of competition. Harvard Business Review, November/December: 77–90.

Porter, M. E. 2000. Locations, clusters and company strategy. In G.L. Clark, M.A. Feldman and M.S. Gertler (eds), The Oxford Handbook of Economic Geography. Oxford: Oxford University Press, 253–74.

Portes, A. 2001. Introduction: the debates and significance of immigrant transnationalism. Global Networks, 1: 181–93.

Portes, A., Guarnizo, L. and Landolt, P. 1999. The study of transnationalism. Ethnic and Racial Studies, 22: 217–37.

Portes, A., Haller, W. and Guarnizo, L. 2001. Transnational Entrepreneurs. Oxford Transnational Communities Project, Working Paper WPTC-01–05. Oxford University.

Prahalad, C.K. and Doz, Y. 1987. The Multinational Mission. New York: The Free Press.

Pratt, T. 2001. Patent on small yellow bean provokes cries of biopiracy. The New York Times, 20 March.

Probst, G., Büchel, B. and Raub, S. 1998. Knowledge as a strategic resource. In S. von Krogh, J. Roos and D. Kleine (eds), Knowing in Firms. London: Sage, 240–52.

Prusak, L. (ed) 1997. Knowledge in Organizations. Boston: Butterworth-Heinemann.

Philadelphia Suburban Corporation (PSC). 2001. Annual Report. Bryn Mawr, PA.

Punnett, B.J. and Ricks, D.A. 1997. International Business. Second Edition, Cambridge, MA: Blackwell.

Rabinow, P. 1996. Essays on the Anthropology of Reason. Princeton, NJ: Princeton University Press.

Raines, A. 2001. The Cluster Approach and the Dynamics of Regional Policy-Making. Regional and Industrial Policy Papers No. 47. European Policies Research Centre, University of Strathclyde.

Ricupero, R. 2001. Rebuilding confidence in the multilateral trading system. In G. Sampson (ed.), The Role of the World Trade Organization in Global Governance. Tokyo: United Nations University, 37–58.

Rifkin, J. 2000. The Age of Access. Harmondsworth: Penguin.

Roesenfeld, S.A. 1997. Bringing business clusters into the mainstream of economic development. European Planning Studies, 5: 3–23.

Ruggie, J. 1993. Territoriality and beyond. International Organization, 47: 139–74.

Rupert, M. 2000. Ideologies of Globalization. London: Routledge.

Russell, P.L. 1995. The Chiapas Rebellion. Austin, TX: Mexico Resource Center.

Saborio, S. 1992. The Premise and the Promise. New Brunswick: Transactions Publishers.

Sader, F. 2000. Attracting Foreign Direct Investment into Infrastructure. Washington, DC: World Bank.

Sally, R. 1994. Multinational enterprises, political economy and institutional theory. Review of International Political Economy, 1: 161–92.

Sampson, G. (ed.) 2001. The Role of the World Trade Organization in Global Governance. Tokyo: United Nations University.

Sassen, S. 1988. The Mobility of Labor and Capital. Cambridge: Cambridge University Press.

Sassen, S. 1996. Losing Control. New York: Columbia University Press.

Sassen, S. 1998. Hong Kong: strategic site/new frontier. In C. Davidson (ed.), Anyhow. New York: MIT Press/Anyone Corp., 130–7.

Saxenian, A. 1999. Silicon Valley's New Immigrant Entrepreneurs. San Francisco, CA: Public Policy Institute of California.

Schoenberger, E. 1997. The Cultural Crisis of the Firm. Oxford: Blackwell.

Schoenberger, E. 1999. The firm in the region and the region in the firm. In T.J. Barnes, and M.S. Gertler (eds), The New Industrial Geography. London: Routledge, 205–24.

Schoenberger, E. 2001. Interdisciplinarity and social power. Progress in Human Geography, 25: 365–82.

Schumpeter, J. 1934. The Theory of Economic Development. Cambridge, MA: Harvard University Press.

Schurman, R. A. 1996. Chile's new entrepreneurs and the 'economic miracle'. Studies in Comparative International Development, 31: 83–109.

Scott, A.J. 1988a. New Industrial Spaces. London: Pion.

Scott, A.J. 1988b. Metropolis. Berkeley, CA: University of California Press.

Scott, A.J. 1992. The spatial organisation of a labour market. Growth and Change, 23: 94–115.

Scott, A.J. 2000. Economic geography: the great half-century. In G.L. Clark, M.A. Feldman and M.S. Gertler (eds), The Oxford Handbook of Economic Geography. Oxford: Oxford University Press, 18–44.

Scott, A.J. (ed.) 2001. Global City-Regions. New York: Oxford University Press.

Scott, A.J. and Storper, M. 1992. Industrialization and regional development. In M. Storper and A.J. Scott (eds), Pathways to Industrialization and Regional Development. London: Routledge, 3–17.

Sedgwick, E. K. and Frank, A. (eds) 1995. Shame and Its Sisters. Durham, NC: Duke University Press.

Shand, H. 2001. Control and Ownership of GM Technology Paper presented at the International Conference on Trade, Environment and Sustainable Development, Mexico City, February.

Shatz, H.J. and Venables, A.J. 2000. The geography of international investment. In G.L. Clark, M.A. Feldman and M.S. Gertler (eds), The Oxford Handbook of Economic Geography. Oxford: Oxford University Press, 125–45.

Shaver, J.M. 1998. Do foreign-owned and U.S.-owned establishments exhibit the same location pattern in U.S. manufacturing industries? Journal of International Business Studies, 29: 469–92.

Shaw, R. 1999. Reclaiming America. Berkeley, CA: University of California Press.

Sheppard, E. and Barnes, T.J. (eds) 2000. A Companion to Economic Geography. Oxford: Blackwell.

Shin, D.-H. 2001. Structures, Strengths, and Beneficiaries of Entrepreneurial Networks. Center for Global International & Regional Studies Working Paper #2001–1, University of California, Santa Cruz.

Shoch, J. 2001. Trading Blows. Chapel Hill, NC: University of North Carolina.

Sidaway, J.D. and Pryke, M. 2000. The strange geographies of 'emerging markets'. Transactions of the Institute of British Geographers, 25,187–201.

Skeldon, R. 1994. Hong Kong in an international migration system. In R. Skeldon (ed.), Reluctant Exiles?. New York: M.E. Sharpe, 21–51.

Sklair, L. 2001. The Transnational Capitalist Class. Oxford: Blackwell.

Skypala, P. 2002. Pulled from the lair, they are not so scary. Financial Times Money FT, 9 March.

Smart, A. and Smart, J. 1998. Transnational social networks and negotiated identities in interactions between Hong Kong and China. In M.P. Smith and L. Guarnizo (eds), Transnationalism from Below. London: Transaction Publishers, 103–29.

Smith, A. 1893/1776. An Enquiry into the Nature and Causes of the Wealth of Nations. London: Routledge.

Smith, A.M. 2002. Imagining geographies of the new Europe. Political Geography, 21: 647–70.

Smith, M.P. 2001. Transnational Urbanism. Oxford: Blackwell.

Smith, M.P. and Guarnizo, L. (eds) 1998. Transnationalism from Below. London: Transaction Publishers.

Smith, N. 1984. Uneven Development. Oxford: Blackwell.

Sölvell, Ö and Bresman, H. 1997. Local and global forces in the innovation process of the multinational enterprise. In H. Eskelinen (ed.), Regional Specialization and Local Environment. Stockholm: NordREFO.

Stafford, B.M. 1991. Body Criticism. Cambridge, MA: MIT Press.

Stafford, B.M. 1996. Artful Science. Cambridge, MA: MIT Press.

Stafford, B.M. 1999. Visual Analogy. Cambridge, MA: MIT Press.

Standard and Poor's Emerging Markets Indices. 2001. http://www.spglobal.com/indexmainemdb.html, accessed on 17 November 2001.

Standing, G. 2002. Beyond the New Paternalism. London: Verso.

Stonehouse, G., Hamill, J., Campbell, D. and Purdie, T. 2000. Global and Transnational Business. Chichester: John Wiley.

Stopford, J. and Strange, S. 1991. Rival States, Rival Firms. Cambridge: Cambridge University Press.

Stopford, J. and Wells, L.T., Jr. 1972. Managing the Multinational Enterprise. New York: Basic Books.

Storper, M. 1995. The resurgence of regional economies, ten years later. European Urban and Regional Studies, 3: 191–221.

Storper, M. 1997. The Regional World. New York: Guilford.

Story, J. 1999. The Frontiers of Fortune. London: Financial Times/Prentice Hall.

Streeck, W. 1996. Lean production in the German automobile industry. In S. Berger and R. Dore (eds), National Diversity and Global Capitalism. Ithaca, NY: Cornell University Press, 138–70.

Suez Lyonnaise des Eaux. 2001. Website www.suez.fr.

Swyngedouw, E. 1997. Neither global nor local. In K. Cox (ed.), Spaces of Globalization. New York: Guilford Press, 137–66.

Swyngedouw, E. 2000. Authoritarian governance, power, and the politics of rescaling. Environment and Planning D, 18: 63–76.

Taylor, L. 1997. The revival of the liberal creed. World Development, 25: 145–52.

Taylor, M. and Asheim, B.T. 2001. The concept of the firm in economic geography. Economic Geography, 77: 315–28.

Taylor, M. and Thrift, N. (eds) 1982. The Geography of Multinationals. London: Croom Helm.

Taylor, M. and Thrift, N. (eds) 1986. Multinationals and the Restructuring of the World Economy. London: Croom Helm.

Taylor, P.J., Watts, M.J. and Johnston, R.J. 2001. Geography/globalization. GaWC Research Bulletin, No.41, Department of Geography, Loughborough University.

The Economist. 1996. Showing Europe's firms the way. The Economist, 13 July.

The Economist. 2001. 'How to see through walls'. 20 September, Technology Supplement, 6.

Thévenot, L. 2001. Organised complexity: connections and co-ordination and the composition of economic arrangements. European Journal of Social Theory, 4: 405–25.

Thiel, J., Pires, I. and Dudleston, A. 2000. Globalization and the Portuguese textiles and clothing filière in the post-GATT climate. In A. Giunta, A. Lagendijk and A. Pike (eds), Restructuring Industry and Territory. London: Stationery Office, 109–26.

Thomas, D. 1995. Chinese media mogul diversifies his empire. Financial Post, C10.

Thornhill, J. 1994. Foreign speculators raise the stakes. Financial Times, 10 October.

Thornhill, J. 1996. Russia: hot spot or black hole? Financial Times, 23 December.

Thrift, N.J. 1997. The rise of soft capitalism. Cultural Values, 1: 29–57.

Thrift, N.J. 1998. Virtual capitalism. In J. Carrier and D. Miller (eds), Virtualism. Oxford: Berg, 161–86.

Thrift, N.J. 1999. The place of complexity. Theory Culture and Society, 16: 31–70.

Thrift, N.J. 2000. Performing cultures in the new economy. Annals of the Association of American Geographers, 90: 674–92.

Thrift, N.J. 2001. It's the romance not the finance that makes the business worth pursuing. Economy and Society, 30: 412–32.

Thrift, N.J. 2002. Summoning life. In P. Cloke, P. Crang and M. Goodwin (eds), Envisioning Geography. London, Edward Arnold, (forthcoming).

Thrift, N.J. and French, S. 2002. The automatic production of space. Transactions of the Institute of British Geographers, 27 (forthcoming).

Thrift, N.J. and Olds, K. 1996. Refiguring the economic in economic geography. Progress in Human Geography, 20: 311–37.

Tickell, A. and Peck, J.A. 1992. Accumulation, regulation and the geographies of post-Fordism. Progress in Human Geography, 16: 190–218.

Tickell, A.T. and Dicken, P. 1993. The role of inward investment promotion in economic development strategies. Local Economy, 8: 197–208.

Tricks, H. 2000. Mexico makes the grade. Financial Times, 9 March.

Tudor, G. 2000. Roller Coaster. London: Reuters/Pearson Education.

Twomey, M.J. 2001. A Century of Foreign Investment in Mexico. Paper presented at the First Congress of Mexican Economic History, Mexico, October, http://www-personal.umd.umich.edu/~mtwomey/fdi/MexInv.pdf, accessed on 25 March 2002.

UBS Warburg (UBS). 2001. Earnings Review: Philadelphia Suburban. New York, 14 June.

Ul Haq, M. 1996. The Bretton Woods institutions and global governance. In P.B. Kenen (ed.), Managing the World Economy. Washington, DC: Institute for International Economics.

Ulmer, G. 1989. Teletheory. London: Routledge.

UNCTAD. 1993. World Investment Report, 1993. New York: United Nations

UNCTAD. 1994. World Investment Report 1994. New York: United Nations.

UNCTAD. 2001. World Investment Report 2001. New York: United Nations.

UNCTAD. 2002. Trade and Development Report 2002. New York: United Nations.

US Department of Commerce. 2000. Environmental Technologies and Services. In US Industry and Trade Outlook, 2000. Washington, DC: Department of Commerce, 20–1 to 20–20.

US Filter. 2001. Website at www.usfilter.com.

US Water News Online. 1997. Demand surging in Asia's water/wastewater markets. www.uswaternews.com/archives/arcglobal/7demsur10.html. November.

US Water News Online. 2000a. Thames buy makes RWE global water no. 3. www.uswaternews.com/archives/arcglobal/tthabuy11.html. November.

US Water News Online. 2000b. Suez Lyonnaise des Eaux completes acquisition of United Water Resources. www.uswaternews.com/archives/arcsupply/tsuelyo8.html. August.

Valdes, J. G. 1995. Pinochet's Economists. Cambridge: Cambridge University Press.

Van Leenep, Emile. 1996. Comments on 'Managing the World Economy Under the Bretton Woods System'. In P.B. Kenen (ed.), Managing the World Economy. Washington, DC: Institute for International Economics.

Vann, K. and Bowker, G. 2001. Installing the truth of practice. Social Epistemology, 15: 247–62.

Velthuis, O. 1999. The changing relationship between economic sociology and institutional economics. American Journal of Economics and Sociology, 8: 629–49.

Vernon, R. 1966. International investment and international trade in the product cycle. Quarterly Journal of Economics, 80: 190–207.

Vicari, S. and Toniolo, G. 1998. Errors and learning in organizations. In G. von Krogh, J. Roos, and D. Kleine (eds), Knowing in Firms. London: Sage, 204–22.

Vivendi Environnement. 2001a. Documents comptables annuels. (Annual Report), Bulletin des Annonces Legales Obligatoires (BOLA), Paris, 13 April: 4809–4828. Available from Vivendi Environnement website: www.vivendienvironnement-finance.com.

Vivendi Environnement. 2001b. Website: www.vivendienvironnement-finance.com.

Vivendi Enviornnement. 2001c. Presentation of Financial Statements, FY 2000.' Available from Vivendi Environnement website: www.vivendienvironnement-finance.com.

von Hippel, E. 1988. The Sources of Innovation. Oxford: Oxford University Press.

von Krogh, G., Ichijo, K. and Nonaka, I. 2000. Enabling Knowledge Creation. Oxford: Oxford University Press.

Wacquant, L. 1999. How penal common sense comes to Europeans. European Societies, 1: 319–52.

Wade, R. 1990. Governing the Market. Princeton, NJ: Princeton University Press.

Walker, R. 1989. A requiem for corporate geography. Geografiska Annaler B, 71: 43–68.

Wall, L., Christiansen, T. and Orwant, J. 2000. Programming Perl. London: O'Reilly.

Wallach, L. and Sforza, M. 1999. Whose Trade Organization? Washinton, DC: Public Citizen.

Waste Management, Inc., 2001a. Annual Report, 2000. Houston, TX.

Waste Management, Inc. 2001b. 10–K Report. Washington, DC: Securities and Exchange Commission.

Weick, K.E. 2001. Making Sense of the Organization. Oxford: Blackwell.

Wenger, E. 1998. Communities of Practice. Cambridge: Cambridge University Press.

Wenger, E. and Snyder, W.H. 2000. Communities of practice. Harvard Business Review, 78: 139–45.

Wever, K. 1995. Negotiating Competitiveness. Cambridge, MA: Harvard Business School Press.

Whatmore, S. and Thorne, L. 1997. Nourishing networks. In D.J. Goodman and M.J. Watts (eds), Globalizing Food. London: Routledge, 287–304.

Whatmore, S.J. 2002. Hybrid Geographies. London, Sage.

Whitford, J. 2001. The decline of a model? Economy and Society, 30: 38–65.

Whitley, R. D. 1992. Business Systems in East Asia. London: Sage.

Whitley, R. D. 1999. Divergent Capitalisms. Oxford: Oxford University Press.

Williamson, O.E. 1970. Corporate Control and Business Behaviour. New Jersey: Prentice-Hall.

Williamson, O.E. 1975. Markets and Hierarchies. New York: The Free Press.

Wolf, M. 2001. After Argentina. Financial Times, 21 December.

Wolf, M. 2002. Countries still rule the world. Financial Times, 6 February.

Wolff, Alan. 1996. Comments on 'Managing the Trading System'. In P.B. Kenen (ed.), Managing the World Economy. Washington, DC: Institute for International Economics.

World Bank. 2000. Global Development Finance. Washington, DC: World Bank.

World Bank. 2002. Globalization, Growth and Poverty. New York: Oxford University Press.

Yeoh, B. and Chang, T.C. 2001. Globalizing Singapore. Urban Studies, 38: 1025–44.

Yeung, H.W.C. 1994. Critical reviews of geographical perspectives on business organizations and the organization of production. Progress in Human Geography, 18: 460–90.

Yeung, H.W.C. 1998a. Capital, state and space. Transactions of the Institute of British Geographers, 23: 291–309.

Yeung, H.W.C. 1998b. Transnational Corporations and Business Networks. London: Routledge.

Yeung, H.W.C. 1998c. The political economy of transnational corporations. Political Geography, 17: 389–416.

Yeung, H.W.C. 2000a. Organising 'the firm' in industrial geography I. Progress in Human Geography, 24: 301–15.

Yeung, H.W.C. 2000b. The dynamics of Asian business systems in a globalizing era. Review of International Political Economy, 7: 399–433.

Yeung, H.W.C. 2001. Regulating 'the firm' and socio-cultural practices in industrial geography II. Progress in Human Geography, 25: 293–302.

Yeung, H.W.C. 2002a. The limits to globalization theory. Economic Geography, 78: 285–305.

Yeung, H.W.C. 2002b. Entrepreneurship and the Internationalization of Asian Firms. Cheltenham: Edward Elgar.

Yeung, H.W.C. 2003. Practicing new economic geographies. Annals of the Association of American Geographers, 93 (forthcoming).

Yeung, H.W.C., Poon, J. and Perry, M. 2001. Towards a regional strategy. Urban Studies, 38: 157–83.

Yoffie, D. 1993. Beyond Free Trade. Boston, MA: Harvard Business School Press.

Zukin, S. and DiMaggio, P. (eds) 1990. Structures of Capital. Cambridge: Cambridge University Press.

INDEX